Palgrave Studies in Mediating Kinship, Representation, and Difference

Series Editors
May Friedman
Toronto Metropolitan University
Toronto, ON, Canada

Silvia Schultermandl
University of Münster
Münster, Germany

This book series brings together analyses of familial and kin relationships with emerging and new technologies which allow for the creation, maintenance and expansion of family. We use the term "family" as a working truth with a wide range of meanings in an attempt to address the feelings of family belonging across all aspects of social location: ability, age, race, ethnicity, nationality, sexuality, gender identity, body size, social class and beyond. This book series aims to explore phenomena located at the intersection of technologies including those which allow for family creation, migration, communication, reunion and the family as a site of difference. The individual volumes in this series will offer insightful analyses of the representations of these phenomena in media, social media, literature, popular culture and corporeal settings.

This series is currently accepting proposals. To learn more, please follow this link: https://www.uni-muenster.de/Anglistik/Research/Amerikanistik/research/index.html

Francesca Decimo

Lives in Motion

The Transnational Making of Population between Morocco and Italy

Francesca Decimo
Department of Sociology and Social Research
University of Trento
Trento, Italy

ISSN 2752-7352   ISSN 2752-7360 (electronic)
Palgrave Studies in Mediating Kinship, Representation, and Difference
ISBN 978-3-031-65582-1   ISBN 978-3-031-65583-8 (eBook)
https://doi.org/10.1007/978-3-031-65583-8

© The Editor(s) (if applicable) and The Author(s), under exclusive license to Springer Nature
Switzerland AG 2024

This work is subject to copyright. All rights are solely and exclusively licensed by the
Publisher, whether the whole or part of the material is concerned, specifically the rights of
translation, reprinting, reuse of illustrations, recitation, broadcasting, reproduction on
microfilms or in any other physical way, and transmission or information storage and retrieval,
electronic adaptation, computer software, or by similar or dissimilar methodology now
known or hereafter developed.
The use of general descriptive names, registered names, trademarks, service marks, etc. in this
publication does not imply, even in the absence of a specific statement, that such names are
exempt from the relevant protective laws and regulations and therefore free for general use.
The publisher, the authors and the editors are safe to assume that the advice and information
in this book are believed to be true and accurate at the date of publication. Neither the
publisher nor the authors or the editors give a warranty, expressed or implied, with respect
to the material contained herein or for any errors or omissions that may have been made.
The publisher remains neutral with regard to jurisdictional claims in published maps and
institutional affiliations.

Cover illustration: Mohamed Tahdaini/Getty Images

This Palgrave Macmillan imprint is published by the registered company Springer Nature
Switzerland AG.
The registered company address is: Gewerbestrasse 11, 6330 Cham, Switzerland

If disposing of this product, please recycle the paper.

In memory of my father Franco (1932–2020)
And
For Alessandro and our playful life together, dancing to the rhythm of a tressette *match*
For my mother Vittoria and all of my beloved, extended family

ACKNOWLEDGMENTS

This book arises from the willingness of a multitude of migrant women and men to tell their stories and open their houses to me: my main and most profound thanks go to all of them, with the hope that the following pages account for the value and strength of their life courses and efforts. I also thank Alessandra Gribaldo and Serena Piovesan for having accompanied me along my empirical research detour: their contribution to making the fieldwork rich, thick, and stimulating has been precious and invaluable.

My fieldwork has also been extraordinarily enriched by the exchanges I have had the opportunity to experience over time with a large number of outstanding scholars. I am deeply grateful to Rainer Bauböck, Laura Bernardi, Paolo Boccagni, Nancy Foner, Alessandra Gribaldo, Roberto Impicciatore, David Kertzer, Sandro Mezzadra, Nazareno Panichella, Giuseppe Sciortino, and Elisabetta Zontini for their comments on earlier and partial versions of my work. I also thank Rocco Molinari, Livia Elisa Ortensi, and Agnese Vitali for their suggestions regarding the statistical data analysis and Elena Zambelli for a final, careful, and exhaustive reading of the whole manuscript and all her precious comments.

My research has been significantly supported by the chance to spend periods as a visiting professor in the US. I am thankful to the New York University Department of Sociology and to Ann Morning for hosting me during the fall of 2016. Later, I spent the fall of 2022 benefiting from a position as a Visiting Research Scholar at the Advanced Research Collaborative (ARC) at the Graduate Center of CUNY (New York), where I participated in its dense schedule of seminars and enjoyed the chance to immerse myself in a stimulating and diversified scientific environment. I am profoundly grateful to Philip Kasinitz, Nancy Foner and all the ARC fellows who shared with me that terrific period of research, attending my seminar and discussing the core arguments of this book. I also thank Robert Smith and Andrés Besserer Rayas for the coffee and chat moments spent discussing my work during that time.

I am sincerely thankful to Angelina Zontine for polishing and improving my English with high competence always embedded in fondness and fun. My deep gratitude also goes to the many people whose presence here and there has made it possible for me to navigate pleasantly through these years of study and writing: Alessandro, my companion, for always being close to me; my mum Vittoria, and Massimo, Lorenzo, Vittoria Jr., and Giuliana for still being my nest in Naples; Giuseppe, Marques, Ada, Danny, Maurice, and Jorge, my family in New York; Cristina, Marcello, Massimiliano, Marcella, Ruba, and all the dear friends I've had the good fortune to enjoy in Bologna.

My final acknowledgments go to the Fondazione Caritro whose funding made possible the research I present in this book, and to the Department of Sociology and Social Research at the University of Trento, where I have grown scientifically and to which I am proud to continue contributing.

CONTENTS

1 Introduction: Migration, Reproduction, and the Place of Families — 1

Family Transnationalism in Times of Border Enforcement — 3
The Geography of Kinship and Family Reproduction — 7
Marriages, Births, and the Boundaries of Demography — 10
Global Horizons, Intimate Connections, and National Constraints — 12
Charting Lives in Motion: Fieldwork Among Moroccan Families in Italy — 15
Chapter Overview — 19
References — 21

2 Family Matter(s): Marriages, Kinship, and Female Mobility — 31

Matchmaking and Transnational Marriages Between Morocco and Italy — 36
Agreed-Upon Choices: Partner Selection Between Dating and Parent Approval — 41
Female Mobility and the Migratory Process — 46
Conclusion: Marriage, Gender, and the Horizons of Family Migration — 50
References — 55

ix

x CONTENTS

3 Children of Migration: The Transnational Making of Population — 59

Living at an Intense Pace: The Interrelation of Migration, Marriage, and Births — 63

Expected Children and "Gifts of Allah": (Un)Planned Pregnancies — 66

The Family Cycle and Offspring as a Reward — 72

Conclusion: Migration and Generation in Spite of Everything — 77

References — 80

4 Copious Relationships: Intimacy and Belonging in Perilous Times — 85

Subjectivity, Conflict, and Family Adjustment — 88

Orchestrated Households: Intimacy and Family Life Across Migration — 94

Conclusion: Family, Displacement, and the Everyday Politics of Intimacy — 101

References — 105

5 Migration, Reproduction, and the Demography of Citizenship — 107

Moroccan Families Among Others: The Aggregate Effect of Intimate Choices — 108

Enacting High Fertility in a Low-Natality Country: A Controversial Lineage — 113

"Blood" Law and the Boundaries of Belonging in Italy — 119

Conclusion: Transnational Migrations, Generation, and the Demography of Citizenship — 124

References — 127

6 Conclusion: Lives in Motion and Their Future — 131

Marriages, Kinship, and the Transnational Making of Population — 133

Transnationalism and the Circulation of Affective Resources — 134

The Structuring of Reproductive Opportunities and the Assimilationist Creed — 136

References — 137

Index — 139

LIST OF FIGURES

Fig. 3.1   Interrelation of marriage, migration, and first childbirth by age of the interviewed Moroccan women   64

Fig. 3.2   Childbirths of the interviewed Moroccan women by age   73

Fig. 5.1   Trends of non-EU immigration in Italy by the ten main nationalities (1995–2023). (Source: Ministry of Interior, residence permits (Data available on the following websites (last access, July 7, 2024): https://demo.istat.it/tavole/?t=permessi&l=it for the years 1995–2008; https://demo.istat.it/tavole/?l=it&t=noncomunitari for the years 2008–2021; and on Istat (2023a) for the 2023))   109

Fig. 5.2   Sex-ratio trend of Moroccan immigrants in Italy (1995–2023). (Source: Ministry of Interior, residence permits [Data available on the following websites (last access, July 7, 2024): https://demo.istat.it/tavole/?t=permessi&l=it for the years 1995–2008; https://demo.istat.it/tavole/?l=it&t=noncomunitari for the years 2008–2021; and on Istat (2023a) for the 2023])   110

Fig. 5.3   Sex-ratio trend of immigrants in Italy by the 15 main nationalities (2000; 2010; 2023). (Source: Ministry of Interior, residence permits (Data available on the following websites (last access, July 7, 2024): https://demo.istat.it/tavole/?t=permessi&l=it for 2000; https://demo.istat.it/tavole/?l=it&t=noncomunitari for 2010; and on Istat (2023a) for 2023))   111

xii   LIST OF FIGURES

Fig. 5.4   Immigrant adult women-minors' ratio by the 15 main female nationalities in Italy (2021). (Source: Ministry of Interior, residence permits [Data available on the following website (last access, July 7, 2024): https://demo.istat.it/tavole/?l=it&t=no ncomunitari])   113

Fig. 5.5   Percentage of children born in Italy of Italian parents and of one or both foreign parents (1999–2022). (Source: Istat (2023b))   114

Fig. 5.6   Birth and death trends in Italy, 2002–2022. (Source: Births and deaths, Istat (Data available on the following websites (last access, July 7, 2024): https://demo.istat.it/app/?i=FE3&l=it for births; https://demo.istat.it/app/?i=ISM&l=it for deaths))   115

Fig. 5.7   Fertility rates in Italy, Italian and foreign women, 2004–2022. (Source: Fertility rates, Istat (Data available on the following website (last access, July 7, 2024): https://demo.istat.it/app/?i=FE1&l=it))   116

Fig. 5.8   Trends of citizenship rejection rate by applicants' continent of origin (2013–2020). (Source: Ministry of Interior [Data available on the following website (last access, July 7, 2024): http://www.libertaciviliimmigrazione.dlci.interno.gov.it/it/documentazione/statistica/cittadinanza])   124

LIST OF TABLES

Table 5.1	Numbers of children born in Italy of one or both foreign parents by the main 15 non-European nationalities (2022)	112
Table 5.2	Percentages of Italian citizenship acquisition through transmission from parents, lineage (ius sanguinis) or choice (18 y/o) and total numbers of Italian naturalization by the main 20 nationalities of origin (2021)	122
Table 5.3	Italian naturalization rate[a] by the main ten nationalities of origin (2021)	123

CHAPTER 1

Introduction: Migration, Reproduction, and the Place of Families

Abstract This introductory chapter presents the book topic and its research foci on migration and the articulations of family life. My investigation aims to rethink the role migrant families and kinship are deemed to play in times of national boundary enforcement, focusing on household closeness and reproduction understood as interwoven elements that work both as a driver and goal of migrants' trajectories. Specifically, I engage the prevailing notion of transnational family to explore: first, how vital events such as marriages, family formation, household settlement, and births are interrelated with and triggered by mobility between contexts of origin and destination; and second, the evolution of the family cycle across migration, the population dynamic this process entails, and the issues of demography and nationality that it raises. The case study under examination will also be introduced in this chapter, by presenting the fieldwork and qualitative data set, collected among 50 Moroccan families living in Italy, that underpins the analysis. To conclude, an overview of the book's various chapters is presented.

Keywords Family transnationalism • Migration and reproduction • Births and marriages across mobility • Immigration • Generation and citizenship

© The Author(s), under exclusive license to Springer Nature Switzerland AG 2024
F. Decimo, *Lives in Motion*, Palgrave Studies in Mediating Kinship, Representation, and Difference,
https://doi.org/10.1007/978-3-031-65583-8_1

How do people in motion make family? What are the times and places of intimacy and generation when individual existences are shaped by migration and uncertainty? How do affective bonds find sufficient ground to engender domestic realms in times of increasing movement, separation, and forced mobility or immobility?

This book stems from questions about how personal life and household formation take place in our age of controversial globalization and contested mobility. Anyone who has crossed paths with someone from elsewhere knows how relevant these issues are, and the answers are equally complex. Indeed, staying close to their own family is a difficult accomplishment for large numbers of mobile subjects whose only means for maintaining contact with their relatives is remote channels. This condition is further exacerbated by economic and legal constraints that may prolong the state of living apart indefinitely, as demonstrated by the many dramatic chronicles of migrants separated from their children by restrictive immigration and border enforcement policies.

It was exactly with this scenario in mind that, as my research questions took shape, I was all the more struck by stories of families in motion that developed in the completely opposite direction. Keenly aware of the multitude of migrants for whom love and family life are widely transmuted into relationships at a distance, I was interested in retracing affective articulations aimed instead at embedding and embracing familial presence within movement. I refer particularly to migratory trajectories that, instead of exposing single individuals to the experience of mobility and displacement, are based on household movement and settlement abroad, showcasing stories of family and generation that evolve by intersecting with mobility. By charting the trajectories of household formation unfolding across migration, I realized that these trajectories were woven by individuals who have been able to dislocate, away from home, the chance of setting up a home, founding spheres of personal care, reproduction, and generation elsewhere. And there is more: I also recognized that precisely through this multitude of personal choices regarding family relationships in motion, these migrants have been able not only to domesticate mobility, as it were, but also to set in play a wider, full-fledged population dynamic across the spaces and boundaries of transnationalism. Indeed, it is through a myriad of choices aimed at entangling affective relations with movement that the demography of nations undergoes significant social and cultural transformations.

Marriages and fertility events lie at the core of this interrelation of family trajectories and national composition, particularly when they imply the birthing and raising of offspring across migration. Viewed in this perspective, bringing children into the world represents a radical event and quite a challenge for mobile subjects who move on uncertain terrain *par excellence*: indeed, births in a foreign land give rise to crucial issues of intimacy, reproduction, and belonging spreading out from the couple formation, family settlement, and household security to views about the future, identity, and kinship, up to issues of parent and children's legal status, nationhood, and citizenship.

This book examines the intimate family lives of migrants and the population dynamics that spring from their personal choices. To do so, it reconstructs life stories and processes of family formation from the standpoint of Moroccan immigrants in Italy, the protagonists of my research. By delving into their *lives in motion*, the following analysis maps how transnational relationships across the Mediterranean have developed over time to support a transformation of this migratory flow from largely single males, migrating alone, into the settlement of households with children. In this vein, my line of reasoning aims to offer a new view onto the transnationalism and reproduction nexus: it considers how migration is intertwined with the evolution of the family cycle, shaping households, and their demography well beyond the sending countries. As I show, not only does care circulate—through material and moral exchanges—amidst transnational families, but vital events such as marriages, family formation, household settlement, and births are interrelated with and triggered by migration.

Family Transnationalism in Times of Border Enforcement

The spatial articulation of migrants' family lives is a topic that has acquired increasing relevance in migration studies, fueled by multiple interrelated strands of analysis. A large number of scholars have inquired into the specific conditions under which, despite spatial separation, family relationships are able to persevere and resist as opposed to weakening or crumbling. This is the research domain in which the notion of "transnational family" was originally forged, a notion understood as the array of individuals connected through social networks, communication flows, practices, values, and feelings that enable them to maintain interpersonal relationships at a distance, sharing a common feeling of family belonging (Bryceson & Vuorela, 2002).

Transnational households have been adopted as an excellent research field to scrutinize the transformation of the private sphere, how gender and generational roles change, and the way the moral, emotional dimension of the family life spectrum evolves under circumstances of separation, distance, and absence (Baldassar et al., 2007; Baldassar & Merla, 2014; Carling et al., 2012; Cienfuegos-Illanes et al., 2023; Goldring, 2004; Kilkey & Palenga-Möllenbeck, 2016; Mazzucato & Schans, 2011). In order to better grasp these processes, Baldassar and Merla (2014) propose the notion of "care circulation" to capture material and affective resources as these are transmitted and received through multidirectional transnational family networks. Similarly, Cole and Groes (2016) define "affective circuits" as the myriad exchanges of material and emotive elements that enmesh goods, people, ideas, and money, bonding those who stay and those who migrate in a shared system of belonging and mutual obligations. Having deconstructed the implicit Western assumption of the household as nuclear and one-locus based, research has thus detected the multiple ways of making family that have existed historically and that are made even more perceptible by contemporary migration (Baldassar et al., 2014; Coe, 2011). In this vein, scholars (Baldassar & Merla, 2014; Cole & Groes, 2016) argue that emotional support and social reproduction may be effectively ensured without the necessity for spatial closeness, underlining that the spheres of intimacy, affectivity, and well-being operate even from afar. Such processes have become relatively rapid in a moment of globalization by digital tech, social media, hyper-connectivity, and more affordable transportation that make family life more sustainable at a distance in a wide range of cases (Foner, 1997). The entire family life cycle can be deployed in such a decentralized space, giving rise to binational as well as multi-national households that orchestrate life courses, reproductive needs, and the exchange of material and moral resources by navigating through different family care regimes, welfare systems, and immigration policies (Bryceson, 2019; Kilkey et al., 2018; Kilkey & Palenga-Möllenbeck, 2016).

Transnationalism indubitably represents a paradigm of reference in migration studies, and family life and mobility issues have been exhaustively scrutinized in light of this theoretical framework. The more I searched for answers to my research questions on the way people on the move make family, however, the more I recognized that a significant part of the picture was being overlooked as long as research focused mainly on the way mobile individuals maintain relationships at a distance. I realized that this approach only partially captures the unpredictable and uncertain ways that individuals and households come to terms with mobility, emplacement, reproduction, and identity. Proximity, attendance, and homemaking, in fact, continue to constitute critical dimensions that, far from being dissolved by transnationalism, require the concrete participation of family members in a specific place, even while their existences are anchored elsewhere and a common understanding of where and what is "home" is anything but obvious (Boccagni, 2016; Fog Olwig, 2002). This is all the more true considering the intense weight that restrictive migratory policies are exerting in forging and impeding movement. More and more frequently, the contemporary enforcement of borders on a global scale and consequent increasing conditions of involuntary immobility plunge the lives of migrants and their families into a state of separation and suspension (Menjívar, 2006; Menjívar & Abrego, 2012). Hence, questioning who and what moves in a transnational family regime gives rise to quite a variegated picture involving the busy circulation of goods, communication, and information and the more controversial and limited mobility of people. As several studies have highlighted, indeed, transnational living represents one of the most harshly segmented conditions of our times: economic and legal constraints substantially curb back-and-forth movements (Bélanger & Silvey, 2020; Cienfuegos-Illanes & Brandhorst, 2023), widening the gap between mobile and immobile subjects depending on their citizenship, class, and race.

A more precise framing for such ambivalences would be "regimes of mobility" (Glick Schiller & Salazar, 2013; Shamir, 2005; Turner, 2007, 2010), highlighting the different degrees of self-determination affecting the condition of being a mobile/immobile subject and investigating not only movement in all its variations but also problematizing its absence. Immobility has indeed long been overshadowed by the celebration of a postmodern era of flows and circulation, represented as indiscriminately inherent to objects, capital, or people. In the face of these processes, even the scholars who argued most convincingly for the potential of transnational care now recognize that nationalistic rhetoric, related restrictive migratory politics, and the consequent stigmatization of "illegal" movements negatively impact not only the chances of disadvantaged immigrants to be mobile subjects but also their well-being and that of their families at a distance (Merla et al., 2020, 2021; Brandhorst et al., 2020).

And yet, there is more. Framing global transnational families through the lens of the mobility regime paradigm also implies critically redefining the relationship between transnational care and technology. Along with celebratory visions of living in a mobility era, indeed, the view of globalization as a force of hyper-connection and participation has been bolstered by the extraordinary achievements made possible by digitalization, datafication, and artificial intelligence. Ever more flawless and widespread communication and AI devices are deeply entangling humans with machines in the intimate and minute routines of daily life. This is the context in which a binary definition of care and communication understood through the proximity versus distance opposition has been questioned, fostering studies that explore the opportunities for co-presence and family transnationalism fueled by the full potential of ICTs (Baldassar et al., 2016). On closer examination, however, the question of who runs what through digital technologies proves much more complicated, especially in migration issues (Leurs & Ponzanesi, 2024; Ponzanesi & Leurs 2022). Ethnographic research on digital devices and the way these are adopted by migrants and their distant families reveals how such technologies can also invade the private sphere in intrusive, ubiquitous ways, so much so as to function as a panopticon (Horst, 2006; Madianou, 2016). At the same time, relational arrangements with people who are present without being "there", as effectively enabled by today's ICTs, imply subtle new asymmetries and introduce paradoxical inversions of closeness and distance: co-present caregivers risk being devalued in comparison to those who remain at a distance but hold on to the power of remittances and decision-making, thereby altering the customary, local "hierarchy of care access" (Acedera & Yeoh, 2021).

Yet, it is precisely the nexus of immobility and technology that is increasingly capturing scholars' attention. Research seeks to make sense of the way the lockdown of societies and remote communication have been entwined and experienced in the wake of the COVID-19 pandemic, on the one side, and the current bordering politics that are increasingly and pervasively implemented by the adoption of biometric technologies of identification and control, on the other side (Tazzioli, 2020). In this scenario, theorizing that transnational care can be sustained by relying on ICT opportunities implies tying the experience of intimacy much more closely to virtual presence than to bodily proximity. In circumstances such as this, daily life is carried out indefinitely under conditions of separation and never-ending waiting: it becomes clear, therefore, that in current times, the inequalities in migrants' family lives manifest more through physical than digital divides.

THE GEOGRAPHY OF KINSHIP AND FAMILY REPRODUCTION

Faced with the contradictory scenario depicted so far, the family transnationalism nexus requires a rethinking in terms of its research scope and questions. Significant insights into such rethinking are offered by a research strand shifting the attention from households themselves to also consider the wider geography of kinship and the role it plays in shaping routes and opportunities of migration (Andrikopoulos, 2023). Adopting this perspective, research is not limited to mapping the range of existing family relationships and how these are deployed at a distance; it also considers the further personal relationships that mobile subjects weave across migration by appealing to relatives of varying degrees. Viewed from this perspective, family can be seen as part of a broader transnational network based on kinship through which mobility is made possible, closeness is pursued, and new households may be established in place.

On the one hand, local and transnational kins constitute the nodes of networks that can be navigated in the search for information, resources, and support, enabling aspiring migrants to create relationships and routes for mobility (Massey et al., 1993; Massey et al., 1994; Massey & España, 1987). Migrants may turn to brothers and sisters, cousins, parents, aunts, and uncles, whether actual or fictive relatives, to ask for money, secure shelter, find work, or access documents (even borrowing the other person's documents), possibly presenting themselves as a partner or arranging a marriage, real or pretend, to obtain the legal right to move and enter the country where they aim to settle down (Andrikopoulos, 2023; Belloni, 2019; Bertolani, 2017; Piot, 2019; Rossi, 2017). Adopting this field of inquiry, scholars highlight how kinship can be adapted, manipulated, and subverted to equip individuals with the most well-suited relationships and attributes for circumventing border controls and falling within the right state's categories of identity and belonging (Andrikopoulos, 2023; Decimo & Gribaldo, 2017).

On the other hand, when family and kinship drive and shape transnationalism, much more than the mere mobility of individuals is triggered. Much has been written about what constitutes family as part of a timely debate aimed at overcoming the notion of kinship both as a biological relationship and as a corporate group (Carsten, 2004; Sahlins, 2013). Some scholars are making a significant analytical effort to transcend the representation of kinship as a social structure, the one prevailing in classical anthropological studies,[1] and pursue investigations of relatedness as "the indigenous idioms of being related" (Carsten, 2000, p. 4), questioning the way relatedness is engendered and made meaningful to the point of being experienced as "natural" affiliation. In this perspective, kinship—whether based on procreation, social construction or both—is understood in terms of shared life, focusing attention on the tiny rituals of daily life to reconstruct the "local imaginary of hearth and home" (Carsten, 2004, p. 40) and those delimited forms of "mutuality of being" (Sahlins, 2013) through which individuals participate materially and morally in each other's existences.

Migration and displacement challenge even the wisdom emerging from recent mappings of kinship meanings and practices, however. Indeed, the key insights of both reciprocity (Sahlins, 2013) and indigenousness (Carsten, 2004) prove slippery points of departure when family and kin matters take place on the move. Mobility is the human experience that blurs precisely that presumed sense of place and local belonging, making it more difficult for studies to link the understanding of kinship to a spatial context of reference—and, even more so, to the house understood as family locus, as suggested by Carsten (2004). On the other side, as stressed by Andrikopoulos (2023), even the interpretation of kinship as mutuality overlooks tensions, ambivalences, conflicts, and violence as regular features running through family life and kinship. This is eloquently illustrated by ethnographic research describing not only stories of mobility that succeed through the kinship support networks but also instances of misunderstanding about mutual obligations as well as the loneliness of individuals whose movement was based on a false assumption of support from relatives, particularly exacerbated in times of economic crisis and legal vulnerability (Andrikopoulos, 2023; Belloni, 2016; Menjívar, 2000).

Each significance and practice of family life across migration needs to be unearthed, navigating through uncertainties of belonging and ambivalences of meanings: these premises show just how relevant the research questions underlying this book are, questions about how home, family, care, and generation are made possible on unstable ground (Maunaguru, 2019, 2021). In particular, I point out that migration changes the way

---

[1] For more on this debate see Carsten (2004, pp. 1–30).

family formation and kinship affiliations are conceived, longed-for, fostered, and achieved as local experiences. The very understanding of love, desire, intimacy, and pairing, as well as the desire to settle down and marry someone with whom to have and raise children—all of these vital events must indeed be imagined and accomplished over time and transnationally, bearing in mind migration as an available option. As my research alongside a rich stream of other studies makes evident, this option is pursued through, and in turn, creates distant kin relationships.

At the same time, the involvement of relatives in governing mobility also affects the social regulation of life courses, the horizons in which they evolve, and the cornerstones through which personhood is shaped and legitimated (Groes, 2016). In this landscape, migration constitutes a socially recognized and highly valorized event, so much so that it is regularly embedded in the way adulthood is imagined and fulfilled (Kleinman, 2016; Kringelbach, 2016; Rossi, 2017; Vidal, 2011). This means that the entire evolution of the family cycle can be restructured as part of the migratory process, and vice versa: viewed in this way, placement and relatedness can be recognized as building on each other. Such a perspective on personhood, kinship, and migration inevitably raises issues of social reproduction. More precisely, it brings to the fore the way individuals, households, and collectivities manage to take certain steps to secure not only their short-term subsistence but also their long-term continuance through generations and across space. Framed through these lenses, the notion of social reproduction I employ throughout my analysis engages with the array of social practices, norms, and dispositions people enact to guarantee care and perpetuate existence both in the everyday present and in view of the future, including procreation as the biological premise of intergenerational transmission (Laslett & Brenner, 1989; Ginsburg Rapp, 1991; Kofman, 2012). Family transnationalism as a paradigm thus encompasses not only family relations at a distance but also closeness and the attainment of new meanings and practices of locality and reproduction. Settlement and the forging of spheres of care and generation in immigration countries are, indeed, highly valued stakes, uncertain, and difficult chapters of family life that migrants nevertheless tenaciously pursue by appealing to and fostering kinship in all its possible global connections.

Marriages, Births, and the Boundaries of Demography

With the aim of delving into the migration-reproduction nexus, *Lives in motion* maps family formation and generation as a process that can be spatially articulated, even including places and contexts far away from the local, original ones. Marriages and births and the way they interrelate with movement are my research foci and the core issues to which this book is dedicated. At the same time, this perspective sheds light on the generative effects of family migration, since marriage makes descendants and kinship which in turn give rise to other subjects, new relationships, and further networks. Such a perspective is quite different from merely focusing on the regeneration of belonging and the creative affective circuits that migrants and non-migrants activate from afar (Cole & Groes, 2016).

Cross-border and transnational marriages play a significant role in the broader process of family transnationalism and the way affective circuits are forged between people and places, de facto representing one of the most significant dimensions through which globalization is concretely experienced and embodied by subjects (Beck & Beck-Gernsheim, 2010; Beck-Gernsheim, 2007). While cross-border marriages generally include all marital bonds in which at least one of the partners is a migrant (Brettel, 2017; Williams, 2010), transnational marriages more precisely result from the transnational networking of kin and personal acquaintances. Specifically, transnational marriages are understood as conjugal couples formed by matching migrants with a spouse coming or descending from the same country of origin, connected through interpersonal community ties (Beck-Gernsheim, 2007; Charsley, 2012). Transnational marriages can be more or less arranged. They can be sponsored by the parents, who suggest or induce a possible match between their son or daughter and a suitable candidate, or they can be pursued by the migrants themselves by commissioning their acquaintances and relatives to search for the right partner on their behalf. On the other side, transnational marriages can also be the result of individuals' choices instead of being arranged, possibly involving the families once the decision to wed has already been made (Pande, 2021).

Indian and Pakistani people in the UK (Ballard, 1990; Charsley & Shaw, 2006; Pande, 2021; Shaw, 2001; Shaw & Charsley, 2006; Werbner, 1990) as well as Turks and Moroccans in Europe (Decimo, 2021, 2022; Lievens, 1999; Timmerman et al., 2009; Van Zantvliet et al., 2014) are among the minorities most exhaustively investigated in relation to this kind of conjugal union. As these studies show, transnational marriages are made possible by personal relationships and kinship politics that extend beyond local spheres of belonging. The relationships elicited to arrange a transnational marriage can span the context of origin and distant places, including all the possible connections that may be delineated by a migratory network at the global level. These ties facilitate the matching of couples and the formation of families destined to settle abroad. They operate at a distance to create spheres of proximity elsewhere, consolidating compasses of care, social control, and reproduction away from the homeland. To grasp such dynamics, researchers overturn the usual perspective focusing on how migrants support their households in their countries of origin to instead consider the quantity of material and immaterial resources that are deployed by families in the country of origin to shape, affect, and support migrants' trajectories and personal choices abroad (Shaw & Charsley, 2006).

Compared to the research strand on marriages, studies on births have been less fully developed in a transnational perspective. Indeed, it is mainly demographers who have explored the migration and fertility nexus (Bohon & Conley, 2015; Kulu, 2005; Kulu et al., 2017; Milewski, 2009), typically adopting national boundaries to delimit the processes under scrutiny and dissecting how migrants and their descendants assimilate to the domestic family norms. Basically, classical studies on population have adopted the national fertility rate as their parameter for comparison with migrants' reproductive behaviors, searching for convergence or divergence and possible explanations.[2] This perspective, driven by uncritically absorbed postulates about modernization and rationality (Johnson-Hanks, 2008), inevitably falls into the trap of methodological nationalism (Wimmer & Glick Schiller, 2002).

A different approach is advanced by Bledsoe et al. (2007) who map the strategies of mobility and reproduction that Gambian migrants enact between their country of origin and Spain, considering their marriage and fertility choices. A crucial element in their movements, indeed, is the chains of conjugal reunification and birth that enhance their chances of settling down in Europe despite adverse migratory policies. The fieldwork conducted by these scholars, notwithstanding its being based in a European country, charts the interrelation of family and mobility choices that migrants weave by projecting their existences toward a transnational

---

[2] See Decimo (2021, pp. 293–294) for a more exhaustive review of these studies.

horizon. In so doing, this investigation brings together migration research with an anthropological wealth of studies (Bledsoe, 2004) showing how different reproductive behaviors are consistent with specific historical and cultural conjunctures, in terms of both the local logics of political economy (Greenhalgh, 1995; Kertzer & Fricke, 1997) and repertoires of meaning linked to intimacy, sexuality, and respectability (Johnson-Hanks, 2002; Schneider & Schneider, 1991).

My investigation adopts this research landscape with the aim of expanding—in a transnational perspective—the body of literature looking at demographic processes as collective vital events embedded in specific arrays of relationships, practices, and meanings (Kertzer, 2005). This means scrutinizing the cumulative transformations driven by the interrelation of migration and family events and generated through processes that take place through the crossing of national boundaries. My analysis contributes to this debate by delving into what I define as the *transnational making of population*, namely an array of relations, practices, and values that make marriage, childbearing, and the raising of offspring sustainable and attractive in migration. This is precisely the interpersonal and spatial dynamics that my approach seeks to comprehend by focusing on couple formation, migrants' chances of having children in the context of immigration, and the ways families from the country of origin maneuver in pursuit of these goals.

GLOBAL HORIZONS, INTIMATE CONNECTIONS, AND NATIONAL CONSTRAINTS

The migratory process as it drives family formation and the making of population constitutes a research field involving not only mobility, borders, and settlement but also issues of identity, nation, and belonging. Family represents a historical target and favored object of interest for the project of nation-building and consolidation (Turner, 2008). It is from the intimacy of the domestic realm, in fact, that the social construction of a population—understood as a nation's primary resource, comprising its body and living organism (Foucault, 1978)—sets out. In this view, households are the main political recipients of biopolitics and the quintessential social arena of governmentality according to which kinship relationships are established and recognized, the distinction between proper and improper sexuality is internalized, the venues of affective and sexual life are naturalized, and procreation is legitimated (Foucault, 2007, p. 85). As such, family is the cornerstone through which gender and generational roles are outlined and performed and on which the very reproduction of the nation is based. Family thus represents a private domain of public interest that is not left up to chance but is instead constantly monitored, supported, limited, or reinforced by the state and its various apparatuses according to the national policies in question (Donzelot, 1979).

Understood in these terms, family is a device of the nation that is expected to intersect with and imbue established class, gender, and ethnoracial constructions. The way nationhood is built, in terms of both imagined community (Anderson, 1991) and as a redistributive regime of mutual welfare (Wimmer, 2002), is based on the distinction between insiders (who behave in a correct way, have economic dignity, display appropriate features, and so on) and those lying outside these parameters (Anthias & Yuval, 1992; De Genova, 2002; Silverstein, 2005). This distinction can be blurred or hidden within the folds of private life and challenged by individual choices, but the categories it engenders are so co-constitutive in the process of identity formation that they insinuate themselves deeply into subjectivity and intimate relationships (Decimo & Gribaldo, 2017). In a similar scenario, families in motion in particular represent the unexpected constituents of an unforeseen demographic evolution. As long as the migratory process moves intimacies and interpersonal bonds, as long as it is enacted by people who articulate families, constitute domestic spheres, and generate populations, it poses thorny questions about the boundaries of the nation and its future. This is the background giving rise to debates about who belongs to the national body, who reproduces it, who is considered legitimately authorized to grant biological and intergenerational perpetuation to its population, and whose contribution to this end is unwelcome (Decimo, 2015; Lonergan, 2024; Moret et al., 2021; Yuval, 1997).

Viewed through these lenses, family can be seen to represent a political matter and migrant families in particular come up against multiple vectors of inclusion and exclusion through which their belonging is judged and debated (Yuval, 1997). Their income and degree of economic independence, their notions of individual autonomy, the gender and generational role they perform, the way they conceptualize family relationships, personhood, affective bonds, and intimate life: all these dimensions are scrutinized by politics and public discourses as signs of belonging or, in contrast, indicators of unwelcome backwardness. An emerging research strand adopting this perspective has insightfully untangled the way intimacy and belonging form a specific political focus through which notions of proper and improper conceptions of family and interpersonal relationships are constructed (Bonizzoni, 2018; Bonjour & de Hart, 2013; D'Aoust, 2022) and the way this operates to constrain and jeopardize migrant families (Bonjour & Cleton, 2021). Delving into laws on family migration and reunification, this scholarship shows that such legislation has become more and more stringent especially for third-country nationals and that questions of cultural integration are increasingly influential in shaping the rules of family-member admission (Andrikopoulos, 2021; Block, 2015, 2021; Jashari et al., 2021; Pellander, 2021; Strasser et al., 2009; Wray et al., 2021). Identity-based criteria and imagined ethnic boundaries can be seen to operate among the courts, administrators, officers, and clerks tasked with categorizing and differentiating between foreign households (Odasso, 2021; Orsini et al., 2021; Ruffer, 2011), thereby threatening migrants' and minorities' personal lives and calling their familial arrangements into question.

Ethnographic research conducted among Sub-Saharan African families in Europe reveals how different conceptions of marriage and parent-child relations enshrined in laws can threaten families with dissolution, particularly dramatically when parents' bonds with their children and their very status as parents are called into doubt (Coe, 2016; Feldman-Savelsberg, 2016). Similarly, many families in the United States have been severely impacted by the recent, further hardening of immigration policies (Massey, 2020) as they are sundered by juridical barriers and indefinitely prevented from staying together. The research has focused in particular on Latino immigrants, with scholars (Abrego, 2014; Castañeda, 2020; Dreby, 2010, 2015; Menjívar et al., 2018) highlighting the "spillover effects" (Aranda et al., 2014) that the border and bureaucracy enforcement regime imply in terms of social, psychological, and even physical harm to migrants and their families. As Boehm (2012) makes evident considering immigration from Mexico, families struggle to co-reside with their children, experiencing the shifting ground of their residence across the border not only as stressful but also as a threat to the well-being and safety of their offspring. In this context, the nuclear, one-locus family appears to be not simply a Western standard but also a condition the yearning for which is fueled by forbidden mobility and prolonged separation (Boehm, 2011, 2012).

In light of this picture, it is clear not only that migrant households emerge as highly relevant political subjects in our current times but also that the broader processes of family formation, reunification, and existence they enact or are impeded from enacting across borders raise questions of reproductive rights. My analysis on social reproduction across migration thus engages with the notion of "stratified reproduction" (Colen, 1995; Ginsburg & Rapp, 1995, p. 3): specifically, conceiving of family reproduction as a highly hierarchized arena[3] intersected by power relations that privilege certain reproductive positions over others, thereby defining who is able to marry and who cannot, who can have children and who cannot, who is worthy of being born and who is not, as well as the practices, resources, and conditions that should apply and the social position they should occupy. My research aim is to delve into the mobility and reproduction nexus by exploring the relationships, practices, and meaning through which the Moroccan migrants, the protagonists of my research, have carved out and consolidated a family domain in Italy: how were they able to establish households and have children despite the adversity of the migratory process? How have they succeeded in this goal in a context such as Italy where marriage, and especially birth, rates are relentlessly decreasing? To what extent are the intimate choices and private spheres of these migrants intersected and affected by the national project? How do their family behaviors challenge the stratified structure of reproduction that underlies the making of Italian socio-demographic features and the way these features are represented?

CHARTING LIVES IN MOTION: FIELDWORK AMONG MOROCCAN FAMILIES IN ITALY

My research experience among Moroccan families extends back to several years ago when I started conducting my fieldwork among migrant women (Decimo, 2005) with the aim to explore how their experience of mobility took shape in Italy. At the time, my main focus was mapping individual patterns of mobility. However, the more deeply I delved into different women's experiences, the more I realized the relevance that family relationships hold in shaping their mobility trajectories. And yet, while a

---

[3] The understanding of social reproduction as a segmented process and its global dynamic of exploitation has been specifically explored by feminist scholars such as Anderson (2000); Aulenbacher et al. (2018); Duffy (2005); Ehrenreich and Hochschild (2003); Glenn (1992); Hochschild (2000); Kofman (2012, 2014); Lutz (2018); Parreñas (2000, 2005, 2012); Sassen (2000, 2003); Tronto (2002); Truong (1996).

wealth of research has examined women's commitment to maintaining their households at a distance, I found the literature to be more limited when it came to problematizing how women move with and settle their families across migration. These insights elicited new research questions regarding the processes enabling migrants to construct and experience a family sphere across mobility. On the other hand, this point was entangled with the demographic evolution brought about by immigration during the 2000s in Italy, a country historically affected by population aging and low fertility levels (Kohler et al., 2002). In this context, demographers observed that foreign-born women have been making a significant contribution, maintaining such a significantly higher fertility rate than Italian women as to raise the national total fertility rate and slightly decelerate population decline (Dalla Zuanna, 2006)—as I discuss later in the book.

It was this array of research experiences, insight and circumstances that inspired the study I present here. Specifically, my analysis is based on ethnographic data collected between October 2013 and November 2014 with 50 Moroccan families with children living in the city of Bologna and the province of Trento, Italy.[4] At the time, these geographical areas hosted the highest immigrant natality and fertility in the country, respectively, with high numbers of Moroccan families constituting significant sites in which to base my fieldwork. The Moroccan immigration flow represented an intriguing research context, given its large numbers and the elevated presence of families. I drew up the research design and approached this fieldwork with the aim of contributing to the debate on migration and fertility from a social science perspective, specifically sociological and anthropological insights into migration studies. However, it was the more recent political climate that granted my research materials a further significant resonance and added a new sense of urgency. The idea for this book was then stimulated by the heated political connotations increasingly used to frame issues of family migration and reproduction, identity and belonging, as well as national population and majority-minority society in Italy and in Europe, accompanied by even stricter politics of national boundary enforcement. Compared with the analyses of a few years ago that welcomed immigrants' contribution to Italian natality as desirable population replacement in a context of lowest-low fertility (Dalla Zuanna, 2006), I was struck by current right and far-right political campaigns increasingly

[4] The research was based at the Department of Sociology and social Research, University of Trento (Italy) and financially supported by the Fondazione Caritro (ref. 2012.308).

hinged on conspiracy theories driven by the racist idea of "ethnic replacement". Propaganda promoting a specifically white, Catholic national identity based on Italian lineage (Morning & Maneri, 2022; Zincone, 2006, 2010) has become particularly insistent even though immigration is by now a long-term structural component of the population (Colombo & Dalla-Zuanna, 2019). This is why I developed this book project, in the conviction that the stories and voices collected among Moroccan families through my fieldwork are more relevant and significant than ever in light of the historical and political moment Italian and European societies are currently experiencing.

I opted to site my research among migrants from one selected place of origin so as to facilitate the social networking that underpinned the empirical construction of the fieldwork: as a result, Moroccan origin did not represent in any way a variable per se in my analysis (Wimmer & Glick Schiller, 2002). Families have been invited to participate in the research in such a way as to diversify the range of different voices while avoiding the risk of redundancy or the uncontrolled selectivity of research subjects. The choice to differentiate between the two research sites was likewise made according to this rationale and not as a basis for comparing them to each other. Furthermore, a variety of channels were used to contact the families, thus limiting my use of the snowball sampling method. The fieldwork was based in multiple districts in Bologna and municipalities in Trentino with the support of a heterogeneity of brokers, both Italian and foreign (public office workers, religious leaders, associations, social workers, teachers, intercultural moderators, and volunteers) who enabled us to make contact with the households involved in the research.

The families involved in the study had between one and six children: a minority of them (14) had four, five, or six children, and the majority (24) had two or three children, while the couples who had one child (12), especially the younger ones, considered additional births to be a possibility. Given this data, the analysis has dedicated particular attention to reproductive events, from marriage to births and their order.

The inquiry was carried out through recorded, unstructured, and in-depth interviews conducted in Italian; however, the fieldwork also involved several moments of observation and ethnographic note-taking. Interviews and observations were conducted by myself in collaboration with two research assistants, with the team meeting periodically to discuss impressions, findings, adjustments, and opinions about the progress and contents of the inquiry. Since one of the topics of the research was pregnancies and childbirths, the interviews were performed mainly with the wives, involving the husbands as well when they were available; in a few cases, when the wives were absent, the interview was conducted only with the husbands.

The interview set took place mainly in the families' homes, in the context of broader encounters that typically lasted two hours or more: before the interview, indeed, it was customary to socialize with neighbors and relatives, drinking mint tea, eating *chebakia, briwat*, and *kaap el ghazal* (Moroccan pastries), and chatting about life events both small and grand; at the same time, the children were often around, and the conversation typically took place while also looking after them and generally participating in those few hours of their everyday domestic lives. In several circumstances, spontaneous focus groups arose, composed of adults and teens hanging out at the house where we gathered, revolving around subjects they themselves brought up and considered relevant to discuss. Therefore, the ethnographic surrounding of the interviews was particularly thick with information.

Before recording each interview, the research was described in detail, laying out the topic, aim, and methods. Specific attention was dedicated to explaining the goals of the study, clarifying that the main purpose was to explore how families have been formed across migration, the obstacles and resources they encounter along their trajectories, and how they manage their daily household lives. Oral consent was provided before the formal outset of each interview.[5]

The fieldwork was also described as a possible space for participants to "have their say" in a context such as Italy where migrant families, especially large and Muslim ones, are more and more frequently subject to judgment and targeted by stigmatizing public discourse. The ethnographic setting allowed us to talk about the interviewees' choice to have children, a choice that they often asserted with pride given their awareness of the low birth rates Italy has been registering for decades. I usually talked with participants about my own position as a child-free woman, and this friendly comparison fostered conversations about pregnancies, personal prospects, and family commitment: we agreed that I was the one who needed to

---

[5] At the time of fieldwork, the ethical rules of the University of Trento—my workplace and the institutional setting for the research—did not require written consent. Concurrently, in compliance with these rules, I explained that the research data would be handled in such a way as to protect them from unintended, unwanted, or malevolent use.

1 INTRODUCTION: MIGRATION, REPRODUCTION, AND THE PLACE... 19

learn something about issues like these and not the other way around. As a white, Italian woman raised Catholic but agnostic since my teens, I was also welcomed in their homes as a witness of their private family lives, as a guest with whom to complain about the absurdities of small and large episodes of everyday racism. On my side, the fact that I come from southern Italy, an area that is itself historically and intensely racialized, made me a particularly sympathetic listener.

The communication thus took place empathically, respecting interviewees' sensibilities and narrative flow more than an abstract or procedural schedule of questions. No specific ethical concerns surfaced during the interviews, thanks largely to careful pre-interview set-up work carried out as part of well-informed relationships with my counterparts, as described above. At the same time, the climate of trust and welcome was furthered by the assurance that the participants' privacy would be protected by anonymizing all names—the ones used in this book are pseudonyms—and modifying or removing any details that might identify them.

The data analysis was conducted by employing a biographical approach (Bertaux, 1981; Bertaux & Kohli, 1984; Chamberlayne et al., 2000; Plummer, 2001; Bornat, 2004, 2008). No specific, pre-determined questions were posed, although the narrative was structured in order to cover key biographical themes (Plummer, 2001; Bornat, 2008), specifically regarding the interviewees' family of origin, couple formation and settlement in Italy, birth events, gender roles, and family relationships and organization. The interviews, fully transcribed taking note of non-verbal expressions and any events that occurred during the conversations, have been read and re-read, cross-compared, and linked in thematic analysis. Several passages from the interviews were selected as key texts in the interpretative process and will be presented in the following empirical sections. The understanding I propose of the way these couples make family across migration has thus been guided by an analysis of individual trajectories and the way each life story informed the comprehension of others, allowing a common interpretative framework to emerge.

CHAPTER OVERVIEW

The research issues delineated so far run throughout the different chapters of the book by first bringing to the forefront the experiences, voices, and worlds of meaning these Moroccan migrants convey and build in their effort to make family in Italy. Only at the end of this detailed exploration does the book step back from these *lives in motion* to reconstruct the political and legal context that surrounds, frames, and constrains their evolution. Specifically, the nexus between marriage and mobility is the focus of the first empirical chapter—Chap. 2—which explores the role played by couple formation in the evolution of migration between Morocco and Italy. Here, I reconstruct the marriage system that arose over time between the two sides of the Mediterranean, the role that kinship plays in facilitating it, and the increasing participation of women in the migratory venture that this process implies. At the end of the chapter, I lay out conclusions theoretically interwoven with the broader debate on transnational marriages, highlighting how these kinds of conjugal formations expand our understanding of the ways families and kinship articulate distance and proximity across migration. Chapter 3 points to the central argument of the book regarding *the transnational making of population*. Here, I focus on the intersection of birth events and the migration process, showing how mobility triggers instead of jeopardizing fertility choices. Basing the analysis on women's life courses, childbirth order, and how reproductive decisions have been made, the analysis looks into the issues of family planning and the value attributed to children at different stages of the family cycle. I close my reasoning on migration and fertility by engaging in a dialog with other ethnographic studies that discuss issues of family norms and reproductive behavior. Then, with Chap. 4, the analysis explores the affective relationships cultivated by individuals and the way subjectivity and intimacy are shaped in the family sphere. I consider, on the one side, circumstances of individual crises and interpersonal conflict regarding the marriage choice or conjugal failure; on the other side, I explore how love and intimacy are represented and experienced in daily family life, considering how gender and generational roles interplay. In the end, I discuss my findings by suggesting an understanding of the affective domain that couples pursue as embedded in an array of relationships, practices, and meanings that materially and morally nurture the accomplishment of their daily lives and reproductive efforts. The final chapter of the book, Chap. 5, shift the focus of the analysis from lived family experiences to their aggregate effect, offering an overall quantitative view of the evolution of the Moroccan presence in Italy among other numerically significant migratory flows. This overview is conducted adopting a bird's eye to sketch how the *transnational making of population* lands in Italian demography and is involved in discourses and policies of national identity and belonging. Most importantly, this chapter argues that the demographic evolution in Italy, as it has unfolded over time also including the immigrant component, does not remotely overlap with the boundaries of nationality: these instead remain fixed rigidly on the principles of "blood" citizenship. Each of the chapters poses final considerations in dialog with the debates relevant to these different topics so as to develop in the conclusive part, Chap. 6, a focus on three main nexuses stemming from my analysis, respectively: marriage, fertility, and kinship; transnationalism and family making; and reproductive resources and the way they are structured across migration.

References

Abrego, L. J. (2014). *Sacrificing families. Navigating laws, labor and love across borders.* Stanford University Press.

Acedera, K. F., & Yeoh, B. S. (2021). When care is near and far: Care triangles and the mediated spaces of mobile phones among Filipino transnational families. *Geoforum, 121,* 181–191.

Anderson, B. (1991). *Imagined communities: Reflections on the origin and spread of nationalism.* Verso.

Anderson, B. (2000). *Doing the dirty work? The global politics of domestic labor.* Zed Books.

Andrikopoulos, A. (2021). Love, money and papers in the affective circuits of cross-border marriages. *Journal of Ethnic and Migration Studies, 47*(2), 343–360.

Andrikopoulos, A. (2023). *Argonauts of West Africa: Unauthorized migration and Kinship dynamics in a changing Europe.* University of Chicago Press.

Anthias, F., & Yuval, D. N. (1992). *Racialized boundaries. Race, nation, gender, colour and class, and the antiracist struggle.* Routledge.

Aranda, E., Menjívar, C., & Donato, K. M. (2014). The spillover consequences of an enforcement-first U.S. immigration regime. *American Behavioral Scientist, 58*(13), 1687–1695.

Aulenbacher, B., Lutz, H., & Riegraf, B. (2018). Introduction: Towards a global sociology of care and care work. *Current Sociology., 66*(4), 495–502.

Baldassar, L., Baldock, C., & Wilding, R. (2007). *Families caring across borders: Migration, ageing and transnational care giving.* Palgrave Macmillan.

Baldassar, L., Kilkey, M., Merla, L., & Wilding, R. (2014). Transnational families. In J. Treas, J. Scott, & M. Richards (Eds.), *The Wiley-Blackwell companion to the sociology of families* (pp. 155–175). Wiley-Blackwell.

Baldassar, L., & Merla, L. (2014). *Transnational families, migration and the circulation of care.* Routledge.

Baldassar, L., Nedelcu, M., Merla, L., & Wilding, R. (2016). ICT-based co-presence in transnational families and communities. *Global Networks, 16,* 133–144.

Ballard, R. (1990). Migration and kinship: The differential effect of marriage rules on the processes of Punjabi migration to Britain. In C. Clarke, C. Peach, & S. Vertovec (Eds.), *South Asians overseas* (pp. 219–249). Cambridge University Press.

Beck, U., & Beck-Gernsheim, E. (2010). Passage to hope. *Journal of Family Theory & Review, 2*(4), 401–414.

Beck-Gernsheim, E. (2007). Transnational lives, transnational marriages. *Global Networks, 7,* 271–288.

Bélanger, D., & Silvey, R. (2020). An im/mobility turn. *Journal of Ethnic and Migration Studies, 46*(16), 3423–3440.

Belloni, M. (2016). 'My uncle cannot say "no" if I reach Libya': Unpacking the social dynamics of border-crossing among Eritreans heading to Europe. *Human Geography, 9*(2), 47–56.

Belloni, M. (2019). *The big gamble: The migration of Eritreans to Europe* (p. 243). University of California Press.

Bertaux, D. (Ed.). (1981). *Biography and society.* Sage.

Bertaux, D., & Kohli, M. (1984). The life story approach: A continental view. *Annual Review of Sociology, 10,* 215–237.

Bertolani, B. (2017). Structural restrictions and personal desires: Arranged marriages between Punjab and Italy. In *Boundaries within: Nation, kinship and identity among migrants and minorities* (pp. 161–180). Springer.

Bledsoe, C., Houle, R., & Sow, P. (2007). High fertility Gambians in low fertility Spain: The dynamics of child accumulation across transnational space. *Demographic Research, 16,* 375–412.

Bledsoe, C. H. (2004). Reproduction at the margins: Migration and legitimacy in the New Europe. *Demographic Research, 3,* 87–116.

Block, L. (2015). Regulating membership. *Journal of Family Issues., 36*(11), 1433–1452. https://doi.org/10.1177/0192513X14557493

Block, L. (2021). '(Im-)proper' members with '(im-)proper' families? *Journal of Ethnic and Migration Studies, 47*(2), 379–396.

Boccagni, P. (2016). *Migration and the search for home: Mapping domestic space in migrants' everyday lives.* Palgrave.

Boehm, D. A. (2011). Here/Not Here: Contingent citizenship and transnational Mexican children. In C. Coe, R. R. Reynolds, D. A. Boehm, J. M. Hess, & H. Rae-Espinoza (Eds.), *2011* (pp. 161–173). Vanderbilt University Press.

Boehm, D. A. (2012). Intimate migrations: Gender, family, and illegality among transnational Mexicans. In *Intimate migrations.* New York University Press.

Bohon, S. A., & Conley, M. (2015). *Immigration and population.* Polity Press.

Bonizzoni, P. (2018). Policing the intimate borders of the nation: A review of recent trends in family-related forms of immigration control. *Gendering nationalism: Intersections of nation, gender and sexuality,* 223–239.

Bonjour, S., & Cleton, L. (2021). Co-constructions of family and belonging in the politics of family migration. In *Handbook on the governance and politics of migration* (pp. 161–172). Edward Elgar Publishing.

Bonjour, S. A., & de Hart, B. (2013). A proper wife, a proper marriage: Constructions of "us" and "them" in Dutch family migration policy. *European Journal of Women's Studies, 20*(1), 61–76.

Bornat, J. (2004). Oral history. In C. Seale, G. Gobo, J. F. Gubrium, & D. Silverman (Eds.), *Qualitative research practice* (pp. 34–47). Sage.

Bornat, J. (2008). *Biographical methods.* In P. Alasuutari, L. Bickman, & J. Brannen (Eds.), The Sage handbook of social research methods (pp. 344–356). Sage.

Brandhorst, R., Baldassar, L., & Wilding, R. (2020). Introduction to the special issue: "Transnational Family Care 'On Hold'? Intergenerational Relationships and Obligations in the Context of Immobility Regimes". *Journal of Intergenerational Relationships, 8*(3), 261–280.

Brettel, C. B. (2017). Marriage and migration. *Annual Review of Anthropology, 46,* 81–97.

Bryceson, D. F. (2019). Transnational families negotiating migration and care life cycles across nation-state borders. *Journal of Ethnic and Migration Studies, 45,* 16.

Bryceson, D. F., & Vuorela, U. (2002). *The transnational family.* Berg Press.

Carling, J., Menjívar, C., & Schmalzbauer, L. (2012). Central themes in the study of transnational parenthood. *Journal of Ethnic and Migration Studies, 38*(2), 191–217.

Carsten, J. (Ed.). (2000). *Cultures of relatedness: New approaches to the study of kinship.* Cambridge University Press.

Carsten, J. (2004). *After kinship.* Cambridge University Press.

Castañeda, H. (2020). *Borders of belonging: Struggle and solidarity in mixed-status immigrant families.* Stanford University Press.

Chamberlayne, P., Bornat, J., & Wengraf, T. (2000). *The turn to biographical methods in social science: Comparative issues and examples* (1st ed.). Routledge.

Charsley, K. (2012). *Transnational marriages.* Routledge.

Charsley, K., & Shaw, A. (2006). South Asian transnational marriages in comparative perspective. *Global Networks, 6*(4), 331–344.

Cienfuegos-Illanes, J., & Brandhorst, R. (2023). Transnational families: Entangled inequalities and emerging challenges on the global scale. In *Handbook of Transnational Families Around the World* (pp. 3–18). Springer International Publishing.

Cienfuegos-Illanes, J., Brandhorst, R., & Bryceson, D. F. (Eds.). (2023). *Handbook of transnational families around the world.* Springer International Publishing.

Coe, C. (2011). How children feel about their parent's migration. In C. Coe, R. R. Reynolds, D. A. Boehm, J. M. Hess, & H. Rae-Espinoza (Eds.), *Everyday ruptures: Children, youth, and migration in global perspective* (pp. 97–114). Vanderbilt University Press.

Coe, C. (2016). Translations in kinscripts: Child circulation among Ghanaians abroad. In J. Cole & C. Goes (Eds.), *Affective circuits* (pp. 27–53). Chicago University Press.

Cole, J., & Groes, C. (Eds.). (2016). *Affective circuits*. Chicago University Press.

Colen, S. (1995). Like a mother to them: Stratified reproduction and West Indian childcare workers and employers in New York. In F. Ginsburg & R. Rapp (Eds.), *Conceiving the new world order: The global politics of reproduction* (pp. 78–102). University of California Press.

Colombo, A. D., & Dalla-Zuanna, G. (2019). Immigration Italian style, 1977–2018. *Population and Development Review*, 585–615.

D'Aoust A-M. (ed.) (2022). *Transnational marriage and partner migration: Constellations of security, citizenship, and rights*. Rutgers University Press.

Dalla Zuanna, G. (2006). Population replacement, social mobility and development in Italy in the twentieth century. *Journal of Modern Italian Studies, 11*(2), 188–208.

De Genova, N. (2002). Migrant illegality and deportability in everyday life. *Annual Review of Anthropology, 31*, 419–447.

Decimo, F. (2005). *Quando emigrano le donne. Reti e percorsi femminili della mobilità transnazionale*. Il Mulino.

Decimo, F. (2015). Nation and reproduction: Immigrants and their children in population discourse in Italy. *Nations and Nationalism, 21*(1), 139–161.

Decimo, F. (2021). The transnational making of population: Migration, marriage and fertility between Morocco and Italy. *Journal of International Migration and Integration, 22*(1), 289–310.

Decimo, F. (2022). Copious relationships: Transnational marriages and intimacy among Moroccan couples in Italy. *Journal of Family Studies, 28*(4), 1255–1271.

Decimo, F., & Gribaldo, A. (Eds.). (2017). *Boundaries within: Nation, kinship and identity among migrants and minorities* (Vol. 24). Springer.

Donzelot, J. (1979). *The policing of families*. Pantheon.

Dreby, J. (2010). *Divided by borders: Mexican migrants and their children*. University of California Press.

Dreby, J. (2015). *Everyday illegal: When policies undermine immigrant families*. University of California Press.

Duffy, M. (2005). Reproducing labor inequalities: Challenges for feminists conceptualizing care at the intersections of gender, race, and class. *Gender & Society, 19*(1), 66–82.

Ehrenreich, B., & Hochschild, A. (Eds.). (2003). *Global woman: Nannies, maids, and sex workers in the new economy*. Metropolitan Books.

Feldman-Savelsberg, P. (2016). Forging belonging through children in the Berlin-Cameroonian diaspora. In J. Cole & C. Goes (Eds.), *Affective circuits* (pp. 54–77). Chicago University Press.

Fog Olwig, K. (2002). A wedding in the family: Home making in a global kin network. *Global Networks, 2*, 205–218.

Foner, N. (1997). What's new about transnationalism? *Diaspora: A Journal of Transnational Studies, 6*(3), 355–375.

Foucault, M. (1978). *The history of sexuality: An introduction* (Vol. 1). Vintage Books.

Foucault, M. (2007). *Security, territory, population: Lectures at the Collège de France, 1977–78.* Springer.

Ginsburg, F., & Rapp, R. (1991). The politics of reproduction. *Annual Review of Anthropology, 20*, 311–343.

Ginsburg, F., & Rapp, R. (Eds.). (1995). Introduction. In *Conceiving the new world order.* University of California Press.

Glenn, N. E. (1992). From servitude to service work: Historical continuities in the racial division of paid reproductive labor. *Signs, 18*(1), 1–43.

Glick Schiller, N., & Salazar, N. B. (2013). Regimes of mobility across the globe. *Journal of Ethnic and Migration Studies, 39*(2), 183–200.

Goldring, L. (2004). Family and collective remittances to Mexico: A multi-dimensional typology. *Development and Change, 35*(4), 799–840.

Greenhalgh (Ed.). (1995). *Situating fertility. Anthropology and demographic inquiry.* Cambridge University Press.

Groes, C. (2016). Men come and go, mothers stay: Personhood and resisting marriage among Mozambican women migrating to Europe. In *Affective circuits: African migration to Europe and the pursuit of social regeneration* (pp. 169–197). University of Chicago Press.

Hochschild, A. (2000). Global care chains and emotional surplus value. In W. Hutton & A. Giddens (Eds.), *On the edge: Living with global capitalism* (pp. 130–146). Jonathan Cape.

Horst, H. A. (2006). The blessings and burdens of communication: Cell phones in Jamaican transnational social fields. *Global Networks, 6*(2), 143–159.

Jashari, S., Dahinden, J., & Moret, J. (2021). Alternative spatial hierarchies. *Journal of Ethnic and Migration Studies, 47*(2), 413–429.

Johnson-Hanks, J. (2002). On the Modernity of Traditional Contraception: Time and the Social Context of Fertility, In: *Population and Development Review, 28*, (2), pp. 229–249.

Johnson-Hanks, J. (2008). Demographic Transitions and Modernity, In: *Annual Review of Anthropology, 37*, pp. 301-315.

Kertzer, D. (2005). Anthropological demography. In D. L. Poston & M. Micklin (Eds.), *Handbook of population.* Springer.

Kertzer, D., & Fricke, T. (Eds.). (1997). *Anthropological demography: Toward a new synthesis.* University of Chicago Press.

Kilkey, M., Merla, L., & Baldassar, L. (2018). The social reproductive worlds of migrants. *Journal of Family Studies, 24*(1), 1–4.

Kilkey, M., & Palenga-Möllenbeck, E. (Eds.). (2016). *Family life in an age of migration and mobility: Global perspectives through the life course.* Springer.

Kleinman, J. (2016). From little brother to big somebody: Coming of age at the Gare du Nord. In J. Cole & C. Goes (Eds.), *Affective circuits* (pp. 245–268). Chicago University Press.

Kofman, E. (2012). Rethinking care through social reproduction: Articulating circuits of migration. *Social Politics, 19*, 142–162.

Kofman, E. (2014). Gendered migrations, social reproduction and the household in Europe. *Dialectical Anthropology, 38*, 79–94.

Kohler, H. P., Billari, F., & Ortega, J. A. (2002). The emergence of lowest-low fertility in Europe during the 1990s. *Population Development Review, 28*(4), 641–680.

Kringelbach, H. N. (2016). The paradox of parallel lives: Immigration policy and transnational polygyny between Senegal and France. In C. Coe, R. R. Reynolds, D. A. Boehm, J. M. Hess, H. Rae-Espinoza (Eds.). (2011). *Everyday ruptures: Children, youth, and migration in global perspective* (pp. 146–168). Vanderbilt University Press.

Kulu, H. (2005). Migration and fertility. *European Journal of Population, 21*(1), 51–87.

Kulu, H., Hannemann, T., Pailhé, A., Neels, K., Krapf, S., Gonzàlez-Ferrer, A., & Andersson, G. (2017). Fertility by birth order among the descendants of immigrants in selected European countries. *Population and Development Review, 43*(1), 31–60.

Laslett, B., & Brenner, J. (1989). Gender and social reproduction: Historical perspectives. *Annual Review of Sociology, 15*(1), 381–404.

Leurs, K., & Ponzanesi, S. (Eds.). (2024). *Doing Digital Migration Studies: Theories and Practices of the Everyday.* Amsterdam University Press. https://doi.org/10.2307/jj.11895524

Lievens, J. (1999). Family-forming migration from Turkey and Morocco to Belgium. *International Migration Review, 33*(3), 717–744.

Lonergan, G. (2024). Reproduction and the expanding border: Pregnant migrants as a 'problem'in the 2014 Immigration Act. *Sociology, 58*(1), 140–157.

Lutz, H. (2018). Care migration: The connectivity between care chains, care circulation and transnational social inequality. *Current Sociology., 66*(4), 577–589.

Madianou, M. (2016). Ambient co-presence: Transnational family practices in polymedia environments. *Global Networks, 16*(2), 183–201.

Massey, D. S. (2020). The real crisis at the Mexico-U.S. Border. *Sociological Forum, 35*, 787–805.

Massey, D. S., Arango, J., Hugo, G., Kouaouci, A., Pellegrino, A., & Taylor, J. E. (1993). Theories of international migration: A review and appraisal. *Population and Development Review*, 431–466.

Massey, D. S., & España, F. G. (1987). The social process of international migration. *Science, 237*(4816), 733–738.

Massey, D. S., Goldring, L., & Durand, J. (1994). Continuities in transnational migration: An analysis of nineteen Mexican communities. *American Journal of Sociology, 99*(6), 1492–1533.

Maunaguru, S. (2019). *Marrying for a future: Transnational Sri Lankan Tamil marriages in the shadow of war.* University of Washington Press.

Maunaguru, S. (2021). (Un)certain futures: rhythms and assemblages of transnational Sri Lankan Tamil marriages. In J. Carsten, H. C. Chiu, S. Magee, E. Papadaki, & K. M. Reece (Eds.), *Marriage in past, present and future tense* (pp. 118–139). UCL Press.

Mazzucato, V., & Schans, D. (2011). Transnational families and the well-being of children: Conceptual and methodological challenges. *Journal of Marriage and the Family, 73*(4), 704.

Menjívar, C. (2000). *Fragmented ties: Salvadoran immigrant networks in America.* University of California Press.

Menjívar, C. (2006). Liminal legality: Salvadoran and Guatemalan immigrants' lives in the United States. *American Journal of Sociology, 111*(4), 999–1037.

Menjívar, C., & Abrego, L. J. (2012). 2012 legal violence: Immigration law and the lives of central American immigrants. *American Journal of Sociology, 117*(5), 1380–1142.

Menjívar, C., Morris, J. E., & Rodríguez, N. P. (2018). The ripple effects of deportations in Honduras. *Migration Studies, 6*(1), 120–139.

Merla, L., Kilkey, M., & Baldassar, L. (2020). Introduction to the Special Issue "Transnational care: Families confronting borders". *Journal of Family Research, 32*(3), 393–414.

Merla, L., Kilkey, M., Wilding, R., & Baldassar, L. (2021). Key developments and future prospects in the study of transnational families. *Research Handbook on the Sociology of the Family*, 439–451.

Milewski, N. (2009). *Fertility of immigrants.* Springer.

Moret, J., Andrikopulos, A., & Dahinden, J. (2021). Contesting categories. *Journal of Ethnic and Migration Studies, 47*(2), 325–342.

Morning, A., & Maneri, M. (2022). *An ugly word: Rethinking race in Italy and the United States.* .

Odasso, L. (2021). Family rights-claiming as act of citizenship: An intersectional perspective on the performance of intimate citizenship. *Identities, 28*(1), 74–92.

Orsini, G., Smit, S., Farcy, J. B., & Merla, L. (2021). Institutional racism within the securitization of migration. *Ethnic and Racial Studies.* https://doi.org/1 0.1080/01419870.2021.1878249

Pande, R. (2021). *Learning to love: Arranged marriages and the British Indian diaspora.* Rutgers University Press.

Parreñas, R. S. (2000). Migrant Filipina domestic workers and the international division of reproductive labour. *Gender & Society, 14*(4), 560–580.

Parreñas, R. S. (2005). Long distance intimacy: Class, gender and intergenerational relations between mothers and children in Filipino transnational families. *Global Networks, 5*, 317–336.

Parreñas, R. S. (2012). The reproductive labour of migrant workers. *Global Networks, 12*(2), 269–275.

Pellander, S. (2021). Buy me love. *Journal of Ethnic and Migration Studies, 47*(2), 464–479.

Piot, C. (2019). *The Fixer. Visa lottery chronicles* (p. 224). Duke University Press.

Ponzanesi, S., & Leurs, K. (2022). Digital migration practices and the everyday. *Communication, Culture and Critique, 15*(2), 103–121.

Plummer, K. (2001). *Documents of life 2*. Sage.

Rossi, A. (2017). Male adulthood and 'Self'-legalizing practices among young Moroccan migrants in Turin, Italy. *Boundaries within: Nation, kinship and identity among migrants and minorities*, pp. 139–159.

Ruffer, G. B. (2011). Pushed beyond recognition? *Journal of Ethnic and Migration Studies, 37*(6), 935–951.

Sahlins, M. (2013). *What kinship is-and is not*. University of Chicago Press.

Sassen, S. (2000). Women's burden: Counter-geographies of globalization and the feminization of survival. *Journal on International Affairs, 53*(2), 503–524.

Sassen, S. (2003). Strategic instantiations of gendering in the global economy. In P. Hondagneu-Sotelo (Ed.), *Gender and U.S. immigration: Contemporary trends* (pp. 43–60). California University Press.

Schneider, J. C. e Schneider, P. T. (1991). Sex and Respectability in an Age of Fertility Decline. A Sicilian Case Study, In: *Social Science and Medicine, 33*(8), pp. 885–895.

Shamir, R. (2005). Without borders? Notes on globalization as a mobility regime. *Sociological Theory, 23*(2), 197–217.

Shaw, A. (2001). Kinship, cultural preference and immigration. *Journal of the Royal Anthropological Institute, 7*(2), 315–334.

Shaw, A., & Charsley, K. (2006). Rishtas: Adding emotion to strategy in understanding British Pakistani transnational marriages. *Global Networks, 6*(4), 405–421.

Silverstein, P. A. (2005). Immigrant racialization and the new savage slot: Race, migration and immigration in the New Europe. *Annual Review of Anthropology, 34*, 363–384.

Strasser, E., Kraler, A., Bonjour, S., & Bilger, V. (2009). Doing family. *The History of the Family, 14*(2), 165–176.

Tazzioli, M. (2020). *The making of migration*. SAGE.

Timmerman, C., Lodewyckx, I., & Wets, J. (2009). Marriage at the intersection between tradition and globalization. *The History of the Family, 14*(2), 232–244.

Tronto, J. C. (2002). The "nanny" question in feminism. *Hypatia, 17*(2), 34–51.

Truong, T. D. (1996). Gender, international migration and social reproduction. *Asian and Pacific Migration Journal, 5*(1), 27–52.

Turner, B. S. (2007). The enclave society: Towards a sociology of immobility. *European Journal of Social Theory, 10*(2), 287–304.

Turner, B. S. (2008). Citizenship, reproduction and the state: International marriage and human rights. *Citizenship Studies, 12*(1), 45–54.

Turner, B. S. (2010). Enclosures, enclaves, and entrapment. *Sociological Inquiry, 80*(2), 241–260.

Van Zantvliet, P. I., Kalmijn, M., & Verbakel, E. (2014). Parental involvement in partner choice: The case of Turks and Moroccans in the Netherlands. *European Sociological Review, 30*(3), 387–398.

Vidal, N. E. (2011). The transnationally affected: Spanish state policies and the life-course events of families in North Africa. In C. Coe, R. R. Reynolds, D. A. Boehm, J. M. Hess, & H. Rae-Espinoza (Eds.), *Everyday ruptures: Children, youth, and migration in global perspective* (pp. 174–188). Vanderbilt University Press.

Werbner, P. (1990). *The migration process: Capitals, gifts and offerings among British Pakistanis.* Berg.

Williams, L. (2010). *Global marriage.* Palgrave.

Wimmer, A. (2002). *Nationalistic exclusion and ethnic conflict.* Cambridge University Press.

Wimmer, A., & Glick Schiller, N. (2002). Methodological nationalism and beyond: Nation–state building, migration and the social sciences. *Global Networks, 2*(4), 301–334.

Wray, H., Kofman, E., & Simic, A. (2021). Subversive citizens. *Journal of Ethnic and Migration Studies, 47*(2), 447–463.

Yuval, D. (1997). *Gender & nation.* Sage.

Zincone, G. (2006). *Familismo legale.* Come (non) diventare italiani. Bari: Laterza.

Zincone, G., & Basili, M. (2010). *EUDO Citizenship Observatory.* Country Report: Italy. EUI, Florence: Robert Schuhman Centre of Advanced Studies.

CHAPTER 2

Family Matter(s): Marriages, Kinship, and Female Mobility

Abstract The nexus between marriage and mobility is the focus of this chapter that delves into the role couple-formation plays in the evolution of migration between Morocco and Italy. In particular, my analysis reconstructs the marriage system that has arisen over time between the two sides of the Mediterranean, examining in depth the role played by: (1) marriage in migration, as an opportunity for and route toward international mobility and a step in the process of familial settling-down; (2) kinship and personal networks in supporting the individual's choice of a partner; and (3) women in promoting a shift in Moroccan movements by making possible familial settlement and the formation of spheres of reproduction abroad. In light of these elements, I discuss the distinction between marriage arranged and by choice, arguing that they represent options along a continuum of possibilities rather than a dichotomy, the meaning of which has to be considered in light of the broader family transnationalism they enact.

Keywords Marriage and migration • Marriage and kinship • Female mobility • Arranged marriages • Marriages by choice

© The Author(s), under exclusive license to Springer Nature Switzerland AG 2024
F. Decimo, *Lives in Motion*, Palgrave Studies in Mediating Kinship, Representation, and Difference,
https://doi.org/10.1007/978-3-031-65583-8_2

31

> Hanan is a young Moroccan woman who married Majeed, her cousin (matrilineal cross-cousin marriage) and came with him to live in a town in northern Italy where her husband's family had been settled for some time. Hanan's sister Maryam also immigrated to Italy: she is single and works in another town as a domiciliary caregiver, a job that is quite common among immigrant women in Italy. Hanan and her husband have a child, and at the time of the interview, they had recently moved to their own house after several years spent living with their in-laws' large family, in crowded and uncomfortable conditions. They are glad to finally have the chance to enjoy their own home-based intimacy and appear to be a happy family. Their son, now three years old, has been enrolled in a nursery school, and Hanan is determined to look for a job to help her husband with the household expenses, but not only. Hanan, like her sister, also wants to help their distant family economically: her mother has fallen ill with diabetes and insulin in Morocco is expensive; her father has lost his job; and there are 6 other siblings to care for, some still quite young. Hanan is looking for a job as a maid. She has always worked as a housecleaner, even when she was a little girl in Morocco, and she used to work with her mother before the latter fell ill. Unlike Maryam, however, Hanan is not interested in an in-home caregiver position because she wants to be able to go home to her family after work and look after her son herself.

This biographical sketch from Hanan's family network opens a window onto a multi-stranded family history in which mobility, care, and gender are variously intersected. Even more so than other stories I collected through my fieldwork among Moroccan households in Italy, this one clearly showcases how family and reproduction are very diversely entangled with migration, outlining different personal destinies.

If we focus on the portrait of Hanan by herself, we see a snapshot of a young mother who, as well as enjoying the warmth of a new family nest with her child, is determined to work outside the home to both gain economic independence for herself and send remittances to her parents in Morocco. However, it is turning the spotlight on the other two figures in this picture, Maryam and Majeed—Hanan's sister and husband, respectively—that we are able to capture the whole of the movement spanning this scene.

Like Hanan, Maryam also migrated to Italy, but unlike her sister, she moved from Morocco as a single woman with the intention of working and helping her family in need who had stayed in the home country. It is significant to compare the trajectories of the two sisters, thus revealing how displacement and generation configure very different scenarios. On the one side, Maryam took the same path that a large number of immigrant women have taken in Italy, that is, securing a position as *badanti* or live-in caregivers. This kind of position implies that the workers reside together with their employers, leaving neither the room nor sufficient time and resources to set up a family life of their own in the country of immigration.[1] Indeed, Maryam—similarly to many other migrant care workers—monetizes her care and reproductive skills to the fullest by addressing them, on the one side, to unrelated people in their private house in form of care labor conducted around the clock; on the other side, this labor is also transmuted into economic remittances destined for the family of origin who remains at a distance. In other words, Maryam is primarily devoted to guaranteeing the well-being of her parents and siblings: she performs the other's reproduction, positing herself like a weld joint between work and care, gender and generations, and place of origin and destination.

On the opposite side, Hanan's trajectory ends up landing in the same place: family, home, mothering, and the chance to work and be productive. While Maryam's reproductive commitment is carried out at a distance, dislocating care and love elsewhere, Hanan's mobility is embedded in a process of family emplacement which expands the domestic scope away from the homeland. While Maryam's reproductive efforts are addressed to her family of origin *there*, in the country of origin, Hanan's are destined to generate a new family and give birth to another generation *here*, in the context of immigration.

Retracing these female stories and looking for the turning point that makes each one so different, the crux that clearly emerges is marriage and its intersection with migration. At the core of Hanan's mobility path, in fact, lies her conjugal match with Majeed and the family dynamics that engendered this occurrence. Majeed is the son of a pioneer of Moroccan immigration to Italy, a man who moved in the 1980s and then reunified his family here; at the end of its cycle, this family was composed of the

---

[1] The Italian domestic labor regime has been considered a significant case study since the first pivotal studies in the debate on care and globalization (Andall, 2000; Parreñas, 2015). There is, indeed, a wealth of research examining how the Italian welfare system has been remodeled through the increasing diffusion of domiciliary care, based on the recruitment of co-resident female immigrant workers. Regarding this phenomenon, see among the others: Ambrosini (2013); Catanzaro and Colombo (2009); Da Roit (2010); Marchetti (2013, 2022); and Sciortino (2004).

conjugal couple, their five children, and the patrilinear grandmother. Like many other Moroccan youths raised in Italy, Majeed used to spend his summers in his country of origin with his family, maintaining strong relationships with their overseas relatives. In such a context of transnational kinship, when Majeed turned 20, his father suggested to him that a marriage with young Hanan could be arranged. This opportunity was welcomed for several reasons: the young people were already close and willing to form a couple; Hanan was also glad to move to Italy, a step understood by her and the whole family alike as a socio-economic improvement compared with the poor conditions of life in Morocco; her parents appreciated the chance to enable her to migrate safely, within a protective kin environment; and at the same time, they trusted that this kind of conjugal agreement would not jeopardize their relationship with their daughter. Indeed, all these hopes have proven true: Hanan is resolute in her intention to send monetary remittances to her parents and siblings and maintains strong ties with them, strengthened by the holidays she regularly spends at their place in Morocco with her husband and their child.

Essentially, Hanan's migration is interrelated with marriage and prompted by a kinship network that, far from weakening at a distance, actually increases and braids together its links and lineages across the Mediterranean. Through this story, we open a window onto a web of *transnational marriages* which, as introduced in the opening chapter of this book, are conjugal unions formed between individuals of the same nationality or national origins of which at least one is a migrant and who are connected through kinship or community ties. In particular, the wedding vow made between Hanan and Majeed reveals a further possible trait of these unions, that is, their being arranged within kinship and between consanguineous relatives (Shaw, 2001). Their union thus sheds light on the fulfillment of an endogamic marriage to its ultimate extent.

Existing analyses of the ways migrations and the household life cycle are intertwined has mostly pointed to household articulations which, as in the case of Maryam, are based on individuals who move abroad to economically support their families remaining in the home country. In this regard, a wide body of literature has extensively scrutinized the evolution of family life and the material and moral bases of the productive roles that migrants, to a large extent women, adopt to ensure care and future prospects for their beloved ones who remain at a distance.[2] We know significantly less about stories such as Hanan's, that is, lives in which family transnationalism is entangled with marriage, female mobility, and generation. By adopting such a perspective, this chapter aims to delve more deeply into the social arena and array of practices and meanings that underpin the formation of transnational marriages as they have been formed over time between Morocco and Italy. In particular, by considering other cases of marriages beyond that of Hanan and Majed and focusing on matchmaking process and female roles, my analysis aims to probe the range of situations, relationships, and personal positions that underlie these conjugal choices, discussing the difference between arranged marriages and marriages by choice. Specifically, in the following pages, I investigate the role played by: a) the family in supporting the transition of individual life courses toward marriage and household settlement in Italy; and b) women and their culture of migration, through which the horizons of mobility have been profoundly transformed in a gendered perspective. I consider transnational marriages as part of the migration process, exploring how they enable a spatial re-articulation of the productive and reproductive spheres of the family cycle which take place through mobility. In this vein, I consider the formation of such conjugal couples as an avenue that, through the vector of female movement, fosters the formation of spheres of care and generation in contexts of immigration. This perspective will be formulated taking into account how a marriage system arose over time between the two sides of the Mediterranean and, at the same time, how conjugal choices are negotiated by subjects, what room for they have agency, and what role is played by gender. The chapter first discusses a distinction that emerged during the course of this fieldwork between so-called "traditional" and "modern" marriages portrayed in the following first and second sections, respectively. Then particular attention is dedicated to considering female mobility, both spatial and social, and how it has changed the social structure of the Moroccan migratory flow to Italy. I conclude the chapter by positioning my research within wider debate on the topic.

---

[2] Beginning with the groundbreaking study by Parreñas (2015) of Filipino domestic workers in Rome, the relationships that female immigrant care workers in Italy maintain with the families they have left behind has been investigated by several scholars, including: Ambrosini (2013); Boccagni (2012); Bonizzoni (2015); Decimo (2005); Marchetti and Venturini (2014); and Vianello (2013).

Matchmaking and Transnational Marriages Between Morocco and Italy

The migratory flow from Morocco to Italy started during the 1980s,[3] concurrently with a widespread closure to immigration by the northern European nations, and the turning of southern ones from emigration into immigration countries (Castles et al., 2014; King, 2000; Van Mol & de Valk, 2016). As various scholars have shown (Decimo, 2005; Persichetti, 2003; Salih, 2003; Zontini, 2010), the first wave of migrants was constituted predominantly by two main figures of working men: married men migrating for economic reasons, leaving their families in the home country and sending them remittances, and single young men who experienced migration as an individual opportunity to improve their conditions and eventually approach adulthood. Since the late 1980s, however, married men—like Majeed's father—have settled in Italy and brought their families to live with them. Most importantly, the arrival of wives and daughters has woven further connections with female relatives, triggering an increase in the female migratory flow (Decimo, 2005). Many of these migrant women married those men who had migrated as bachelors, thus enabling this first male cohort to marry after migration rather than before and opening up new opportunities of mobility and life course evolution for women.

These are the historical premises for the constitution of transnational networks through which, over time, migration and marriage have become interwoven to an increasing extent for young Moroccans. Men could be bachelors who migrated as singles in search of new opportunities and experiences, or second-generation youths who have been raised here. In both cases, it is through marriage that long-term plans and settlement take shape: they are able to meet the right partner in the context of destination but, especially during the initial period when the demography of Moroccan settlement in Italy was predominantly masculine, the possibility of considering a woman from the country of origin as a potential bride was appreciated, as made evident by the case of Majeed described above.

In many cases, the search for the right mate took place during the summer holidays that these men used to spend in their country of origin. The

---

[3] For a broader picture of Moroccan migration in Europe and its historical evolution, see, among others: Berriane et al. (2015); De Haas (2006, 2007); Ennaji (2014); and Lahlou (2019).

classical, old-fashioned way of locating the right partner is through family and relatives, and these figures are significantly involved in the quest and often the first to suggest the potential bride to the candidate groom. Many of the interviews do indeed refer to having had a "traditional" marriage— enacted by more than half of the interviewees—understood as conjugal formations arranged by parents and kin. The partners may never have met before, as in the case of Lamya: she told me with irony and self-confidence that she agreed to consider a marriage proposal with a cousin she did not know, someone who had migrated to Italy several years before and was presented to her in a photo. Similarly, the following accounts by Nada and Youssef, respectively, clearly explain how families arranged their marriages by choosing the right partner on their behalf. Youssef's story also reveals how a first date might be included in the ritual, to allow the candidates to have the last word on the marriage decision:

> *When did you meet your husband?*
> We got married in 1990
> *And the choice of husband was made by you, or they helped you …*
> No, no, no, family members […] I was young, let's say we had an old-fashioned marriage, like from the 50s–60s … but even now some people do it that way!
> *But did you like him?*
> Yes, of course! Even now, after 22 years of marriage!
> *And the engagement lasted how long?*
> 6 months.
> *Was he in Morocco at that time or was he already in Italy?*
> No, he had been here since the 1980s.
> *After you got married and he was away, did you stay with your parents or did you go to live with his parents?*
> No, with his mom, his family, and he was there too, he came and went, because then I wasn't in Morocco for long, the children did kindergarten here [*they left shortly after the birth of the second child, born just after the first one*] (Nada)

> So, there was a family nearby, they went to a wedding and they saw her … they know that I was looking for a girl to marry because I am here [in Italy] … I can't, I don't have time to be looking for a girl like that. You know, we're a bit … conservative … do you understand what I mean by this? We are different from you, to marry, we have this culture […] so, to look for a girl you have to ask the family, find out about her background, how she was educated, these things are important

Who asks her about these things? Do your parents ask?

No, all of us together. I want to know, too, because in the end I am the one stuck with it (*laughing*). Do you understand what I mean? I'm the party in question! (*laughing*)

So, my mother arrived, she said: well, there's the girl, they told me that [she's] polite, beautiful, so all of us together, let's go and see her first. First, they went and then my mother said: okay, we liked her, you have to see her, too. So, afterwards I went, too. We made an appointment, a day to see her, and we all went together. I saw her, they gave me a room to do the interview ... and I fell in love! (Youssef)

The story of Youssef allows us to retrace all the steps and social networking that took place in Morocco to lead him, a single man working and living in a small, isolated village in northern Italy, to transform his life by setting up a family in the context of immigration. At the same time, both Nada and Youssef stress that attraction and love at first sight are featured in their meetings with their promised spouses. Nada in particular, with her long conjugal story, speaks to how pleasure and fondness are still at the foundation of their life as a couple of decades later.

However, in the majority of the stories of arranged marriages I collected, the spouses already knew each other through kinship or neighborhood relationships, including long-standing ones—as in the case of Hanan and Majeed considered at the beginning of this chapter. The following account from Lubna reconstructs how a marriage can be agreed on within the circles of family relationships, between individuals who are already close:

He is from Casablanca, I am from Rabat, the administrative capital. How did we meet each other? There is a relationship between the families. He came when he was young, I saw him several times and then the proposal. Then his father and mother proposed to me, as happens for so many boys and girls, because we have that thing in our customs and habits, that families can, as it were, enter into marriage. Then the decision came back to me—it's just a proposal. And he talked to me, we talked many times and I saw that I liked him. And then it worked out.

And how long did the engagement last?

Not long, we got to know each other for a year, a year and a half, also by phone, when he came down to Morocco, and also through letters. He had already been in Italy since '90. Even writing letters (Lubna)

2 FAMILY MATTER(S): MARRIAGES, KINSHIP, AND FEMALE MOBILITY 39

Many of these couples marry after a short engagement and settle down in Italy soon after the wedding or, as in the case of Nada, when the children have just been born. Indeed, these women disdain the chance to stay in Morocco with their parents-in-law and play the role of wives at a distance: in their view, marrying someone who lives abroad means having the chance to be migrants themselves.[4] Such a perspective is made clearer by this account from Saalima, whose marriage was suggested by the woman who would become her sister-in-law. As Saalima declares, the idea of making a family and coming to Europe are entwined parts of the same wish:

> It was enough, it's [not necessary] that much time, I saw [him] once, for a month, through his sister. He came to Casablanca, he's been in Italy for many years, he came to visit his sister. He came to visit me, he asked me if I wanted to get married, and so on. I had the idea that I wanted to get married, that it would make me happy, to a gentleman, and then there was the idea of my coming to Europe, of my having a family, and everything else. And I saw that he prayed, my dad saw it too, and plus he is a handsome man, so, in 15 days we decided. Then he went to Italy, we did the documents, [had] the wedding and two days later I came with him here. First, we did the documents and had a family party, then when the visa and the papers to come here with him were ready, we had a big party in Casablanca. (Saalima)

In a nutshell, the Moroccan female repertoire of desires, projects, ambitions and relationships has been increasingly molded in view of the migratory horizon. Women have forged their own culture of female mobility (Decimo, 2005; Salih, 2003) in which marriage plays a significant part. The opportunity to marry and reunify with a migrated man is associated with prestige and good fortune: as Saalima explains, such a choice represents the main way of coming to live in Europe, achieving a well-off social position, and giving their offspring a better future. Furthermore, migration enables women to peacefully avoid the norm of patrilocality and constitute their own nuclear households (Lievens, 1999; Timmerman, 2008; Timmerman et al., 2009) in which relatives, including parents-in-law, can be hosted at need but are not themselves the hosts.

Basically, traditional, arranged marriages are now embedded in a wider social perspective which implies geographical distance between the

---

[4] These preferences are the opposite of those observed by Hannaford (2017) in her fieldwork among transnational Senegalese brides, as these women assert their desire to make a life in their country of origin, staying bonded to their migrant husbands as wives at a distance.

conjugal couple and their kin, that is, the foundation of nuclear households and the increasing legitimation of women as mobile subjects. In this framework, it is significant to consider the cases of unmarried migrant women who move by navigating through family and kinship networks as daughters, sisters, aunts, nieces, and cousins. Similarly to their male counterparts, these women and their parents also tend to look to the country of origin to search for the right mate. This is the case of Faatin, for instance, who allowed her family and kin to arrange a marriage which, in this case, implied the husband's reunification with the wife, as the couple interviewed together explain:

> *Hassan*: Faatin, my wife, is related to my sister-in-law, who lives in the area here in [...], my brother's wife, they are related, although a bit distantly, let's say second degree, and through that family of my sister-in-law, I got to know this family, I used to visit them often ... (*Faatin follows the story, chuckling*). It didn't come about [spontaneously] ... I mean, with us there is always, not really a 'separé', but our way of getting to know a girl is not how other people think of [doing] it, that there is freedom to talk to or date a girl or to go out, absolutely not. We both grew up in very traditional families. [...] My mom at the time took an interest—as we say now, every mother is trying to get her children settled. The first one had already married, and now there was the second one, which would be yours truly. And from there it came about, slowly ...
> *All of this happened in Morocco?*
> *Hassan*: Yes, but she came to Italy before I did. [...] An idea was developed that, from the engagement, then the families basically agreed etc. and then from there the track was laid, it was just a matter of looking for the train (*laughter*) and we found it ...
> *The engagement was short, then?*
> *Faatin*: two weeks!
> *Hassan*: well, but we had known each other for a long time
> *Faatin*: I knew him well, he used to come around all the time ...
> *Hassan*: and as soon as I threw out the bait she said yes, she didn't wait that long ... after a week (*laughter*). After that it's not like we dated or lived together, family-wise, basically, in terms of our parents, you know? And in September of '91 we got married. Again, in Casablanca. It was a September wedding, by early October we were already in Italy, a honeymoon and also a business trip! She left first, then after the wedding we both came, [we were] practically newlyweds. (Faatin and Hassan)

In conclusion, these stories show how the traditional manner of arranging marriages in Morocco has been adapted to a transnational horizon of mobility, thereby engendering a dynamics of family reproduction that is more and more intertwined with migration. Networks of relatives and acquaintances are mobilized for these matchmakings, roundtrips to find the right mate are elicited, and weddings are celebrated, all of which fuel practices and imaginaries that cumulatively innovate and valorize this process. Births, second generations, divorces, and further marriages increase the scope of the transnational arena, fostering social and cultural change and increasing the existential relevance of this global horizon of kinship.

AGREED-UPON CHOICES: PARTNER SELECTION BETWEEN DATING AND PARENT APPROVAL

"Traditional" marriages represent only a section of the wide range of practices through which women and men meet, date, and marry between Morocco and Italy. Among the conjugal stories I collected, almost half narrate encounters and pairings that took place in a much more direct and spontaneous way than found in encounters mediated by the families. These kinds of unions, described as "modern" marriages, are based on the chance to wed someone met independently of relatives, someone with whom the initial approach is driven by appeal and personal involvement. Courtship, typically led by men, is often initiated in the public sphere, where young people can make their first contact in a way that evades the social control dynamics associated with the private domain. On the street, in commercial spaces, or in any other public environment, even making eyes at someone can be an effective hook to start a relationship, as Nahla declares. Her account, along with those of Muhja and Haadiya, depicts the different situations through which their love affairs were triggered. At the same time, these stories clarify how such matches, in order to be freely experienced, need to be grounded in the family sphere: it is in this sphere that such pairings are expected to be introduced and carried forward as a serious commitment, regardless of how they began:

> We are both from Casablanca. He comes from a family that lives below my dad's house, and he used to visit his aunt and then—a glimpse! He had been in France in '86 for three years, then couldn't get papers there, so he went back to Morocco, then to Libya. He is older by 10 years. Then when he went back to Morocco, he found me! (*laughter*)

And how did you manage after the first glimpse?
In Morocco it's not like it is here, that you just pick someone up and take them home. He saw me, asked his aunt, she said I was a good girl. His aunt had gone to dad—I had talked to my mom—I was 18. Then, little by little, love grows. Not like here where they stay engaged for 7 years, even 10! And then later they get married—that's not how we do it! (*laughter*). (Nahla)

Would you tell me about how you and your husband met?
We met, because in Casablanca he lives near my school, so we met at my high school, we talked, he got my address, he came to my house, and everything was fine
How long were you engaged for? A short time?
Yes, a very short time, because he was here [in Italy], he was only coming for 10 days, then he left and came back after one year and we did family reunification (Muhja)

Can you tell me about your husband? How did you two get together?
We have … a modern marriage, not that I didn't know him … he didn't spend time at my house [he was not related to the family] … I met him, then we got engaged
Where did you meet?
At a party! He was singing!!! We met, and from there …
So you involved your families afterwards—but were they happy about it?
Well … at first, my family was not very supportive, in part because he was very young … I was 20, he was 21 at the time, it seemed like we were just playing … like it was a game, not a marriage … then later I convinced them (Haadiya)

These accounts make evident that the approaches through which partners become closer to each other are based on individual agency and personal preferences, but also that the parents' consent is required. Essentially, their relationship needs to be embedded in the family roles and rituals and be legitimized as an engagement quite soon after meeting, in order to proceed without incurring social disapproval. It might take some time to secure their agreement, as described by Haadiya, daughter of an upper-class family from Rabat, but typically the parental involvement gives rise to the betrothal, and this can rapidly progress to marriage. Indeed, the idea of sustaining a long engagement before marrying is widely disvalued, as Nahla proudly declares: waiting to wed is not deemed a sign of wisdom, and intimacy is instead expected to build slowly, step by step, after the marriage and over the course of the conjugal story—as the analysis in Chap. 4 will show in more depth.

Furthermore, it is relevant to stress that no significant differences emerge from a comparison of the ways love stories begin and evolve in the country of origin versus that of immigration. The case of Malika is emblematic in this regard: she arrived in Italy when she was 12 years old; she lived here in her teens and had some previous relationships before meeting the man who would become her husband when she was 22. Like the women's accounts presented above, her story also opens a window onto the way a first encounter may take place in public space, how the boundaries between stranger-ness and intimacy have been drawn, and what criteria she adopted to consider him a proper candidate. Malika was conscious of her needs and the life course phase she was undergoing, and she defined a threshold he had to pass to succeed in his courting, that of introducing himself to her father and declaring his serious intentions, as she clearly explains:

> I was at the market, I alone, he with his friend. I had a cast on my leg and I was on crutches, and I was supposed to go for a checkup in Verona, but I had missed the train, so I thought I would go to the market. He was the one who saw me and said something, like "how is your leg," some words … and I answered him
>
> *You didn't know each other …*
>
> No, no … then I went to the train station, I'm waiting for [the train to] Verona, and he comes with his friend, has coffee at the station bar and then he jokes, he says, "I'll go with you" and I say "yes, yes come on, come" (*laughter*). Then he asks me for my name, surname, and phone number and, after that, he said: "when you come back from Verona call me and I'll give you a ride home"—because I used to live up in the mountains. Later, when I came back, there was no bus, there was nothing, I made the phone call, he came and gave me a ride home and that was it.
>
> But at that time, he was already thinking about wanting to get married, and I was thinking about it too. My mom had been dead for a year and I was thinking about looking … because after my mom died, I didn't think about fooling around anymore. He was thinking that way, too, so he talked with his family by phone, called his mom, said he met a girl well, one with a family like this and that … and she said "okay."
>
> *So how long were you engaged for?*
>
> Not long at all, only 8 months, the time to get Moroccan papers.
>
> *And how did you know he was the right one for you?*
>
> Because I want to marry, I don't want to play around! Because he immediately went to talk to my father …
>
> *So you knew he was the one, because he was serious?*

44   F. DECIMO

> Yes, and good, you know! Because he wasn't like the others—because at 22 I understand everything, you know!!! I had been there before ... (Malika)

Analogously, the following passages of interviews from Sana and Nihad—women who like Malika grew up in Italy—make clear at what stage of the relationship the parents' agreement is required. At the same time, these accounts also describe the rapid and dynamic way in which, once the family is involved, the process can culminate in marriage. Sana, in particular, describes which norms and rules enable partners to date, specifying that while an engagement can be broken at any time if it is not working, a relationship maintained privately, without making the connection public and known, can bring disapproval and blame:

> *How did you meet your husband?*
> He didn't live here [in Italy], he lived in Spain ... but I used to spend time with his mother and sister and every now and then I saw him when he came to them, to visit them. One day he decided to settle here with his parents ... And from that point, we began to see each other
> *And did your parents approve?*
> No ... they didn't know ... we were seeing each other in secret! Because (*laughing*) it's impossible to see each other [publicly]
> *Oh, really?*
> They do not accept the fact that ... they believe that the man must come to ask your parents for your hand. We dated for a short time, we saw that we were getting along, after that he introduced himself
> *So, you are saying that, even just to date, a man should already ask ...*
> Ask for your hand, yes ... then if it doesn't work out and everything goes wrong, the engagement is broken ... so at least people don't say that she is seeing him secretly ... We did it secretly for a little while ... after, we ... we told everyone, he came to ask for my hand, my parents agreed, I agreed, and that's it, we started dating ...
> *How long would you say the engagement lasted?*
> We had only a short ... everything ... everything was quick ... the engagement was ... we got engaged in February 2007 and in August 2007 we were already married (*laughter*)—very short (Sana)

The need for the couple to involve their families and receive their consent is all the more essential when the partners aim to date not only in Italy but also in Morocco. The following account by Nihad explains how roles and behaviors have to change appropriately between one context and the other. Being aware of the different behaviors expected in one scenario versus the other, she pushed her boyfriend to assume a proper position in relation to her family when in Morocco. He amply demonstrates that he was up to the challenge, surprising her with a formal marriage proposal with all the attached ceremonials and rituals required by the occasion:

> Yes, I met him here [...] I met him and my brothers had also seen him. I used to tell them that he was the brother of a friend of mine named Miriam. Then after 6 months I went to Morocco and he also came afterwards and he was like, "ah, I'll see you around." I answered him: "but you know, we are in Morocco ... it's not like we're in Italy! In Morocco you know, I can't go out ... we have, you know, another world." And then I say: "look, if you really love me, if you want me, then you have to introduce yourself to my parents." He is from Casablanca, I am from Khouribga ... we are 120 km apart! So, he brought his family, his parents, and his brothers came there to talk to my parents! So, when my brothers saw him they said: but this guy? Isn't this Miriam's brother? In the end he actually had a sister named Miriam! (*laughter*) But she doesn't live in Italy-she lives in Morocco! (Nihad)

Nihad's story clearly conveys that Italy and Morocco are part of the same transnational arena across which personal relationships and marriage stories are deployed: in such an arena, a twining of family and kin ties underpins the evolution of life courses whether they begin on the one side of the Mediterranean or the other. The custom, quoted in all these interviews, of negotiating the conditions of the marriage with kin in Morocco, or celebrating there as a second step to the wedding already held in Italy, provides evidence of this bidirectional kin networking. Fundamentally, this whole constellation of practices and ties come together to form a single-family system of constraints and opportunities: on the one side, this arena of relationships supports individuals in search of the right mate and fosters the celebration of marriage; on the other side, it works as an effective transnational network of social control which can be manipulated, adjusted, and twisted by individuals but always is taken seriously.

In the end, by retracing how these different encounters developed into conjugal vows, the analysis has brought to the fore the way personal transnational networks underpin the evolution of individual life courses. What emerges is a transnational family-making system that mobilizes social resources with the aim of facilitating and supporting the formation of couples destined to base their homes in Europe. This also means that, despite their differences, in all the stories I collected family and kinship features as far from irrelevant in the way a marriage is entered into, whether through arrangement or personal choice. Alongside this distinction, the accounts I collected also repeatedly described a tapestry of rules, subjectivities, and norms defining the role relatives play in individual decisions. Marriage is indeed not considered a private, individual institution (Bourdieu, 1972, p. 58, 2008) among the people I interviewed, as they instead consider it self-evidently appropriate and reasonable to involve their parents and kin, to varying degrees, in the choice of a spouse. In this framework, the difference delineated so far between "traditional", arranged marriages and "modern", independently chosen ones reflects not a sharp pattern of opposition, but rather various outcomes along the same continuum of possibilities.

Female Mobility and the Migratory Process

Besides the pivotal part that parents, relatives, and transnational networking play in supporting matchmaking and marriages between Morocco and Italy, the core element that makes the evolution of these family movements possible is the participation of women in the migratory process, including the increasing participation of single women (Sadiqui, 2019). Not only do women welcome the chance to marry a migrated man so as to move themselves, as I pointed out above. Moreover, they experience the chance of mobility as an opportunity to work and achieve economic independence in an easier and more lucrative way than in Morocco. In this regard, it is worth mentioning that Moroccan women's employment rates in Italy are rising, even though they are still much lower than those of men.[5] This participation in the labor market represents a remarkable datum considering the difficulties they have in balancing work and household tasks as part of daily routine, unlike migrant women whose families stayed behind in their countries of origin.

Within Moroccan settlement in Italy, female participation in the labor market turns out to be related to the evolution of the household system: far from being hindered, working women are now increasingly considered a productive resource for the domestic economy, so much so that their income is sometimes a matter of contention between the different parties of a marital agreement. Nihad, whose relationships were presented in the

---

[5] See in this regard the report produced by the Italian Ministero del Lavoro e delle Politiche Sociali (2022).

previous pages, explains in this vein how her parents received the marriage proposal from her fiancé, highlighting that her salary represented the stake her father objected to losing:

> *And were your parents happy?*
> Yes, yes, although my father no, he doesn't want [it]
> *Why not?*
> Because I used to work ... I used to bring home a salary ... it was another salary ... because when you get married, your salary doesn't go to your parents anymore ... and you have to think instead about your own home (Nihad)

Women are aware of the need to position themselves at the intersection of diverse needs and spheres of belonging, encompassing the family of origin, their own fulfillment and satisfaction, and the possible new family they might form. The following account by Afaf further elucidates how female agency is enacted by managing different roles. Afaf migrated to Italy as a single woman, but at the same time, she "trained" her parents to accept her freedom by nurturing her relationship with them, not only keeping up a continuous flow of communication but also regularly sending economic remittances. The choice to marry stands out in her story as an autonomous and conscious decision, consistent with the steps she had already made and the results she had achieved:

> I had been living since '90 in Bologna, and I got married in '94. Quite a long time, because this decision that is not a minor one. I didn't want to get married to an Italian, because ... I really care about my customs, my culture. I didn't want to change; and I didn't want to force anybody to change for me. That was always my idea. Even the children, I didn't want to ... I didn't want to have foreign children who would one day reject my culture, my way of doing things, my way of being, my parents' way of being. That was something that I cherished. So, I had a first relationship, but it didn't go well. Then I met my husband, we loved each other and decided to get married.
> *And your parents let you proceed freely?*
> Yes, because I had prepared them, first! (*laughing*) The worry was always there, however, we were always in touch by phone. And I always helped them as much as I could, because every month I was able to send something. I worked like crazy all day long, from 6 in the morning to 9 at night, on the books and off the books. With the cooperative on the books, and then sometimes with the family, I was always able work a few extra hours to round out [my earnings] ...

Didn't your parents say, like, why don't you look for a husband here in Morocco? Did they not propose some guys for you?
I personally am a girl who from a young age had a strong character! It's difficult. The first thing that pushed me to leave Morocco was this, this culture, that they rule over us, they … but that doesn't mean that all Moroccan women are like that (*laughing*). Because even in Morocco, it's not that …
And so, your husband, you found him on your own, let's say, here in Bologna. How old were you when you got married?
We got married in '94. I was 28. He is my age, too.
So, you had a little year of engagement and then you got married.
That's right.
Down in Morocco or here?
No, we did a thing here, at the consulate. Waiting for time off; then we went down and celebrated with family. (Afaf)

Afaf introduces multiple topics regarding the development of female life courses across migration, pointing out how subjectivity evolves and the way women's agency is played out through heterogeneous social expectations and bonds. Her story shows how women are able to move from Morocco as single women. At the same time, Afaf asserts her choice to wed a conational instead of taking a chance with a mixed marriage, revealing that she identifies with a line of belonging linked to her ancestry. The way she shapes her subjectivity and desires is consistent, on the one side, with a female *habitus* (Bourdieu, 1972, 2008) that embodies issues of intergenerational transmission; on the other hand, it illustrates the agency and behavior that she personally reinterprets in adopting transnational horizons of social reproduction. In a nutshell, through a mindful combination of attachment to and detachment from her world of origin, Afaf has proven capable of expanding the space available for her own self-determination.

Analogously to the story of Afaf, Majida and Haniya also relocated from Morocco as single women, going to stay with their uncle and brother, respectively. In a similar way, these women's trajectories and the choices they made are likewise centered on the autonomous economic positions they achieved. It is worth stressing the supportive role played by these male relatives regarding their decisions:

Majida: I studied and I did two years of university, I was studying literature, then I went home, however I had an uncle here and he got me to call, I came here, I did seasonal work and I was here for 9 months. After that I went back to Morocco, and then I got another contract and came back here. Then there was an amnesty and I got all the documents [2006]. After that I worked a lot of jobs, I worked as a caregiver, I worked in a hotel, I worked with a company: I would always work a little bit, then it would end and I had to find something else.

Where were you living at that time? At your uncle's?

Majida: At first, I stayed with my uncle and his family, then I learned the language, got a job, and then also rented a place on my own, an ATAS house [local non-profit migrant advocacy association]. Then through my cousin, my uncle's son, I met him [Rachid]

Rachid: Yes, I was alone, I met her cousin in Italy and told him that I wanted to get married, that I was looking, and so we met, we got along well ...

Majida: yes, he was in Modena and so long distance, we were talking by phone ... we got married after 3 years.

So, your families did not know each other, they met afterwards?

Majida: Yes, they wanted to get to know this man well, first, but then they said it was okay, they were happy (Majida and Rachid)

I came here when I was 19, he [her brother] had told me, "come on, come, take a trip, come to us"—because it was just him with his wife and daughter, they didn't have anyone else in the house ... because I was very attached to this brother of mine, who is older and who was a second father to me. I came as a tourist to stay three months and then I was supposed to come back but instead he really convinced me to stay. There were [training] courses at the time for hairdressers, cooks ... I liked being a hairdresser then, but when I went to register 100 of us showed up and they only accepted 25! I didn't speak Italian well, so my brother told me, "listen, they definitely won't accept you here, better start the course as a cook." I started it, but after a week I got the letter for the hairdressing course, I had been accepted! (Haniya)

Haniya ended up training for two years, at the end of which she was hired by one of the more prominent salons in town. She was autonomously living her life in Italy when, some years later, she met the man she would go on to marry in a shopping mall. She took some time to verify that his "Southern mentality" fit with her own, one definitely based on self-determination as she explains in continuing her story:

We met in Bologna, yes, he is from Morocco, and I am from Morocco ... we were in a shopping mall, he had asked me something, he heard me talking on the phone, he heard my accent, so he came to talk to me ... because he cared, he really wanted someone from Morocco! So, we left the mall, I had to go to the butcher, he suggested a place for me to go, because we only get meat from our [Halal] butcher, so we went the same way. So, we met here and became friends, we used to talk on the phone a lot and we used to go out together ... although at the beginning, with his mentality, I wouldn't have said it could work ... because he's from the south while I'm from the north, we have two different mentalities, anyway, and at the beginning I struggled. Then later I became fond of my husband and then we ... fell in love and then later ... (*laughing*) married! We had a little party, we went down, I introduced the family to him, I mean he came with his family and everything ... (Haniya)

As the sequence of events reconstructed by Afaf, Majida, and Haniya clearly conveys, working and being independent were the priority aims in their migratory trajectories; the chance to marry followed from rather than anticipating the fulfillment of this goal. Through the dissemination of a female culture of mobility, indeed, women's paths have been decoupled from the roles of wives and mothers, shifting more and more into positions based on work and autonomy. At the same time, these working women enable the foundation of dual-income households equipped to deal with the economic challenges immigrant families must face. In the end, this gender dynamic took shape through a more general transformation of family norms and customs that encompasses and complements female trajectories of enfranchisement instead of denying and rejecting them. On these bases, marriage still represents a significant event women pursue in their life courses, but they do so from positions of independence and with meanings and practices that, through migration, contribute to enhancing their autonomy.

CONCLUSION: MARRIAGE, GENDER, AND THE HORIZONS OF FAMILY MIGRATION

Marriage is acquiring increasing relevance in terms of driving global migration. While a great deal of attention has long been paid to framing mobility as a process molded by economic and labor market forces, research has made it ever clearer that family formation and the evolution of this cycle on a global scale represent a crucial dimension that underpins the migratory process (Bailey & Boyle, 2004; Cooke, 2008; Kofman, 2004). In this process, cross-border marriages constitute a pivotal link through which individuals and households significantly shape the routes and horizons of migration (Beck & Beck-Gernsheim, 2010; Williams, 2010; Brettel, 2017).

The number of individuals who move worldwide by marrying someone with a different national or residency status is soaring, pointing to new fields of inquiry and debate about the way mobility and intimacy are interrelated. Dating, pairing, intimacy, marriages, divorces, and re-marriages that take place through mobility do indeed delineate what Constable (2005) defines as "marriage-scapes", coming to represent a new, translocal geography of family formation that unfolds by crossing borders and interweaving conjugal relationships on a global scale. The breadth of the research agenda on this topic gives an idea of the thick social processes involved. It ranges from considering the processes of intimacy-commodification underlying the global spread of cross-border marriages (Constable, 2009) to questioning how subjectivity, desires and gender roles come into play to orient the choice of a distant spouse, as well as figuring out how assorted forms of couple-formation shape the demography of minorities and their assimilation or segregation in the long run (Alba, 2020; Vasquez-Tokos, 2017). In this landscape, individuals—and particularly women—are variously portrayed along a continuum from "victim" to "agent", depending on the different arrays of resources or constraints that are entangled in cross-border marriages (Yeung & Mu, 2020).

However, a further cogent perspective emerges when research focuses on the specific features of transnational marriages such as those depicted by my analysis here. The fieldwork I conducted among Moroccan families in Italy reveals a family-making system that mobilizes social resources to support the formation of couples destined to be based abroad, in Europe. Three pivotal points can be seen to underpin this system and its effectiveness across the times and spaces of transnational migration: (1) Marriage is not considered a private, individual institution. After the first moment of contact, the potential couple is strongly encouraged to secure parental and family consensus to move ahead, even if they come together by choice and independently of any family influence. Given these premises, representing marriages by choice as if they were somehow opposed to arranged marriages reiterates a false dichotomy. This observation is similar to Pande's findings regarding marriages among the British Indian population (Pande, 2015, 2016, 2021); (2) Directly or indirectly, personal transnational relations play an active role in sponsoring transnational marriages: by connecting individual migrants to kin and friends, these relations sustain the formation of new couples and the perpetuation of a family life across migration; (3) women lead this process, since it is only through their determination to move, the roles they come to play throughout the migratory venture, and their will to marry transnationally that such a reproductive domain can be expanded from Morocco to Italy and Europe.

Essentially, kinship and migration constitute a social arena through which transnational marriages take place between the two sides of the Mediterranean. It follows that, from one instance to the next and through a myriad of back and forward movements, migrant networks have been forging a social system that expands and adapts the customary spaces of family cycle evolution to a broader life scenario (Cole & Groes, 2016; Fleischer, 2007). As highlighted here by adopting the gender perspective, such a process has been fostered by the spread of a female culture of mobility that valorizes these patterns of social and spatial mobility as opportunities for women's enfranchisement from patrilocal norms, and consuetudinary, local bonds of social control. As pointed out by studies on the kinship-migration nexus, these transnational movements are embedded in a process of social innovation that takes place by reinventing relatedness, belonging, and personhood. In this perspective, kinship is depicted not as a system of norms, roles, and rules but as a social network that individuals are able to creatively navigate in order to establish identities and relationships that better suit their horizons of mobility (Andrikopoulos, 2023; Cole & Groes, 2016; Decimo & Gribaldo, 2017). This view acquires further significance in light of the increasing fortification of borders which are making mobility dramatically difficult and risky all across the Mediterranean. In these circumstances, migration through marriage represents a privileged pathway for moving safely and ensuring one's status in a foreign country.

These processes also affect the way locality and belonging are perceived and represented when intersected by trajectories of global mobility (Kandel & Massey, 2002; Levitt, 1998, 2001). Several studies conducted in diverse areas of Morocco (De Haas, 2006, 2007; Elliot, 2021; Vacchiano, 2021) underscore that these places of origin have changed as a result of the perpetuation of mobility networks and the proliferation of social and economic remittances they convey (Boccagni & Decimo, 2013). Rural communities, regions, and urban districts, inhabited by high numbers of individuals who experience migration as a habitual feature of their existence, now include the "outside" (Elliot, 2021) as part of ordinary local life. Individuals and households may have direct experience with overseas' lifestyles or indirect contact, based on the economic resources and communication flows circulating in their spheres of belonging. In both circumstances, migration is represented as a regular feature of their social horizon, a possibility for imagining a destiny if not the *only* chance to secure any future at all (Vacchiano, 2021).

Local cultures thus become imbued with values, sentiments, tastes, and desires based on mobility and the lifestyles that take place elsewhere, whether experienced or imagined, as they are deeply entrenched into the repertoire of people's possible life choices. Still, if migration is widely represented as a passage to adulthood and transitional event in the life course (Kandel & Massey, 2002), this is even more true when it is embedded together with marriage in that this interrelation triggers a further mobilization of emotional, affective, and social resources. The couple stories presented so far are clearly engendered by these processes: the straightforwardness, readiness, and effectiveness with which my interviewees met and wed between the two sides of the Mediterranean clearly illustrate the impact that transnationalism has had over time in terms of reconfiguring their arena of reference around marriage. Family norms and customs, as well as individual tastes, preferences, and desires related to pairing, conjugality, and love, are now shaped in keeping with wider spaces and new frontiers of biographical evolution.

However, the marriage stories retraced here also show that these dynamics develop in a specular, reciprocal way *here* and *there*: on the one side, through relatedness as a process that renders the local fascinated by and intimately linked to the global; on the other side, in some respects, foreignness is domesticated at a distance by expanding spheres of family involvement and belonging away from home. In this sense, the array of family rituals, rules, and values that underpin the practice of transnational marriages is part of a wider process of social and cultural adjustment that aims to exercise some kind of control over what takes place abroad. Interpersonal relationships, trust, and reciprocity represent precious collective assets considered by individuals and households when gaging the reliability and value of a possible conjugal match. Cross-border marriage does indeed entail distance and separation and may constitute quite a risky event for the individual in his or her life course, as well as for the well-being of the household left behind. Under these circumstances, it may be appropriate and reasonable to readapt the long-established praxes according to which individual behaviors remain attached to well-known spheres of belonging and social control (Charsley, 2006; Shaw & Charsley, 2006). My analysis has highlighted that several of the couples whose stories I retrace here were formed through the arrangement of marriages by parents (Hense & Schorch, 2013; Pande, 2014) and pairing between consanguineous individuals (Charsley, 2007; Reniers, 2001; Shaw, 2001). In a similar way, other studies report that customs such as polygamy (Charsley & Liversage, 2013) or traditional duties like the dowry (Biao, 2005) and bride price (Bélanger & Linh, 2011) are enhanced or reintroduced to govern transnational marriages. Far from being unwittingly reproduced from generation to generation as archaic traditions, these practices are part of contemporary family strategies through which parents aim to protect and control the destinies of their children when the new conjugal couple settles abroad (Charsley, 2006), enhancing their own social positions in places of origin and avoiding the risk of draining the local context of its "socio-emotional commons" (Isaksen et al., 2008) by sending them off into an unspecified global horizon.

Understood through this lens, transnational marriages constitute a reproductive opportunity stemming from fields of relationships constructed and stratified across time and space. As such, they evoke the notions of habitus and the social capital of marriageability (Bourdieu, 1972, 2008). On the one side, the habitus of love, intimacy, and family-making changes by adopting a global horizon of desire and normalizing migration as the main road for achieving social mobility and a satisfying family life. On the other side, the relationships underlying the formation of these marriages constitute social capital, a capital that is unevenly distributed among migrants of different minorities and statuses. Specifically, the analysis of the Moroccan migrants' conjugal stories I have conducted sheds light on the way this form of capital enables a mobilization of affective and social resources that makes possible the dislocation of households away from the consuetudinary places. This is a process which cannot be taken for granted, considering the widespread diffusion of transnational family patterns based on migrant workers' distance from the households they leave behind, as suggested by the comparison between Hanan's and Maryam's stories that provided the starting point of this analysis. In this perspective, the conjugal formations and family settlements I describe unveil a specific geography of household reproductive and productive spheres that significant numbers of Moroccan families are able to forge by keeping them tightly mutually embedded throughout their movements between the two sides of the Mediterranean. Transnational marriages understood in this way represent not only an act of adjusting family life to a global horizon, but vital links in a broader web of reproductive investment that, by granting participants a future away from home, engenders a population dynamic destined to unfold transnationally.

REFERENCES

Alba, R. (2020). *The great demographic illusion: Majority, minority, and the expanding American mainstream.* Princeton University Press.

Ambrosini, M. (2013). *Irregular migration and invisible welfare.* Palgrave.

Andall, J. (2000). *Gender, migration and domestic service.* Ashgate.

Andrikopoulos, A. (2023). *Argonauts of West Africa: Unauthorized migration and kinship dynamics in a changing Europe.* University of Chicago Press.

Bailey, A., & Boyle, P. (2004). Untying and retying family migration in the New Europe. *Journal of Ethnic and Migration Studies, 30*(2), 229–241.

Beck, U., & Beck-Gernsheim, E. (2010). Passage to hope. *Journal of Family Theory & Review, 2*(4), 401–414.

Bélanger, D., & Linh, T. G. (2011). The impact of transnational migration on gender and marriage in sending communities of Vietnam. *Current Sociology, 59*(1), 59–77.

Berriane, M., de Haas, H., & Natter, K. (2015). Introduction: Revisiting Moroccan migrations. *The Journal of North African Studies, 20*(4), 503–521.

Biao, X. (2005). Gender, dowry and the migration system of Indian information technology professionals. *Indian journal of gender studies, 12*(2–3), 357–380.

Boccagni, P. (2012). Practising motherhood at a distance. *Journal of Ethnic and Migration Studies, 38*(2), 261–277.

Boccagni, P., & Decimo, F. (2013). Mapping social remittances. *Migration Letters, 10*(1), 1.

Bonizzoni, P. (2015). Here or There? Shifting Meanings and Practices in Mother–Child Relationships across Time and Space. *International Migration, 53,* 166–182.

Bourdieu, P. (1972). *Outline of a theory of practice.* Cambridge University Press.

Bourdieu, P. (2008). *The bachelors' ball.* Chicago University Press.

Brettel, C. B. (2017). Marriage and migration. *Annual Review of Anthropology, 46,* 81–97.

Castles, S., De Haas, H., & Miller, M. J. (2014). *The age of migration: International population movements in the modern world.* Palgrave Macmillan.

Catanzaro, R., & Colombo, A. (Eds.). (2009). *Badanti & Co.* Bologna.

Charsley, K. (2006). Risk and ritual: The protection of British Pakistani women in transnational marriage. *Journal of Ethnic and Migration Studies, 32*(7), 1169–1187.

Charsley, K. (2007). Risk, trust, gender and transnational cousin marriage among British Pakistanis. *Ethnic and Racial Studies, 30*(6), 1117–1131.

Charsley, K., & Liversage, A. (2013). Transforming polygamy: Migration, transnationalism and multiple marriages among Muslim minorities. *Global Networks, 13*(1), 60–78.

Cole, J., & Groes, C. (Eds.). (2016). *Affective circuits.* Chicago University Press.

Constable, N. (Ed.). (2005). *Cross-border marriages: Gender and mobility in transnational Asia*. University of Pennsylvania Press.

Constable, N. (2009). The commodification of intimacy. *Annual Review of Anthropology, 38*, 49–64.

Cooke, T. J. (2008). Migration in a family way. *Population Space and Place, 14*, 255–265.

Da Roit, B. (2010). *Strategies of care: Changing elderly Care in Italy and the Netherlands*. Amsterdam University Press.

De Haas, H. (2006). Migration, remittances and regional development in Southern Morocco. *Geoforum, 37*(4), 565–580.

De Haas, H. (2007). Morocco's migration experience: A transitional Perspective. *International Migration, 45*(4), 39–70.

Decimo, F. (2005). *Quando emigrano le donne. Reti e percorsi femminili della mobilità transnazionale*. Il Mulino.

Decimo, F., & Gribaldo, A. (Eds.). (2017). *Boundaries within: Nation, kinship and identity among migrants and minorities* (Vol. 24). Springer.

Elliot, A. (2021). *The outside: Migration as life in Morocco*. Indiana University Press.

Ennaji, M. (2014). *Muslim Moroccan migrants in Europe: Transnational migration in its multiplicity*. Springer.

Fleischer, A. (2007). Family, obligations, and migration: The role of kinship in Cameroon. *Demographic Research, 16*, 413–440.

Hannaford, D. (2017). *Marriage without Borders: Transnational spouses in neoliberal Senegal*. University of Pennsylvania Press.

Hense, A., & Schorch, M. (2013). Arranged marriages as support for intra-ethnic matchmaking? A case study on Muslim migrants in Germany. *International Migration, 51*(2), 104–126.

Isaksen, L. W., Devi, S. U., & Hochschild, A. R. (2008). Global care crisis: A problem of capital, care chain, or commons? *American Behavioral Scientist., 52*(3), 405–425.

Kandel, W., & Massey, D. S. (2002). The culture of Mexican migration: A theoretical and empirical analysis. *Social Forces, 80*(3), 981–1004.

King, R. (2000). Southern Europe in the changing global map of migration. In R. King, G. Lazaridis, & C. Tsardanidis (Eds.), *Eldorado or fortress? Migration in Southern Europe*. Palgrave Macmillan.

Kofman, E. (2004). Family-related migration: A critial review of European Studies. *Journal of Ethnic and Migration Studies, 30*(2), 243–262.

Lahlou, M. (2019). Morocco and the European Union. In M. Ennaji (Ed.), *The Maghreb-Europe paradigm: Migration, gender and cultural dialogue* (pp. 6–23). Cambridge Scholars Publishing.

Levitt, P. (1998). Social remittances: Migration driven local-level forms of cultural diffusion. *International Migration Review, 32*(4), 926–948.

Levitt, P. (2001). *The transnational villagers*. Univ of California Press.

Lievens, J. (1999). Family-forming migration from Turkey and Morocco to Belgium. *International Migration Review, 33*(3), 717–744.

Marchetti, S. (2013). Dreaming circularity? Eastern European women and job sharing in paid home care. *Journal of Immigrant & Refugee Studies, 11*(4), 347–363.

Marchetti, S. (2022). *Migration and domestic work.* Springer.

Marchetti, S. and Venturini, A. (2014). Mothers and Grandmothers on the Move: Labour Mobility and the Household Strategies of Moldovan and Ukrainian Migrant Women in Italy. *International Migration, 52,*111–126.

Ministero del Lavoro e delle Politiche Sociali. (2022). Rapporto Annuale sulla presenza dei migranti: Marocco. Retrieved August, 3, 2024, from https://www.lavoro.gov.it/documenti-e-norme/studi-e-statistiche/rapporto-annuale-sulla-presenza-dei-migranti-2022-marocco

Pande, R. (2014). Geographies of marriage and migration: Arranged marriages and South Asians in Britain. *Geography Compass, 8*(2), 75–86.

Pande, R. (2015). I arranged my own marriage. *Gender, Place & Culture, 22*(2), 172–187.

Pande, R. (2016). Becoming modern. *Social & Cultural Geography, 17*(3), 380–400.

Pande, R. (2021). *Learning to love: Arranged marriages and the British Indian diaspora.* Rutgers University Press.

Parreñas, R. S. (2015 [2001]). *Servants of globalization: Women, migration, and domestic work.* Stanford University Press.

Persichetti, A. (2003). *Tra Marocco e Italia.* CISU.

Reniers, G. (2001). The post-migration survival of traditional marriage patterns: Consanguineous marriages among Turks and Moroccans in Belgium. *Journal of Comparative Family Studies, 32*(1), 21–45.

Sadiqui F. (2019). Women and migration in the context of globalization in in Ennaji, M. (Ed.). The Maghreb-Europe Paradigm: Migration, Gender and Cultural Dialogue. *Cambridge Scholars Publishing,* 44–54.

Salih, R. (2003). *Gender in transnationalism.* Routledge.

Sciortino, G. (2004). Immigration in a Mediterranean welfare state: The Italian experience in comparative perspective. *Journal of Comparative Policy Analysis: Research and Practice, 6*(2), 111–129.

Shaw, A. (2001). Kinship, cultural preference and immigration. *Journal of the Royal Anthropological Institute, 7*(2), 315–334.

Shaw, A., & Charsley, K. (2006). Rishtas: Adding emotion to strategy in understanding British Pakistani transnational marriages. *Global Networks, 6*(4), 405–421.

Timmerman C. (2008). Marriage in a 'Culture of Migration'. Emirdag Marrying into Flanders. *European Review. 16*(4):585–594. https://doi.org/10.1017/S1062798708000367

Timmerman C., Lodewyckx I., Wets J. (2009). Marriage at the intersection between tradition and globalization. *The History of the Family, 14*(2):232–244.

Vacchiano, F. (2021). *Antropologia della dignità. Aspirazioni, moralità e ricerca del benessere nel Marocco contemporaneo.* Ombre corte.

Van Mol, C., & de Valk, H. (2016). Migration and immigrants in Europe: A historical and demographic perspective. In B. Garcés-Mascareñas & R. Penninx (Eds.), *Integration processes and policies in Europe: Contexts, levels and actors (p. 206)* (pp. 31–55). Springer Nature.

Vasquez-Tokos, J. (2017). *Marriage vows and racial choices.* Russell Sage Foundation.

Vianello F. (2013). Ukrainian migrant women's social remittances. *Migration Letters 10*(1), 91–100.

Yeung, W. J., & Mu, Z. (2020). Migration and marriage in Asian contexts. *Journal of Ethnic and Migration Studies, 46*(14), 2863–2879.

Williams, L. (2010). Global Marriage: Cross Border Marriage Migration in Global Context. Plagrave.

Zontini, E. (2010). *Transnational families, migration and gender: Moroccan and Filipino women in Bologna and Barcelona.* Berghahn Books.

CHAPTER 3

Children of Migration: The Transnational Making of Population

Abstract This chapter sketches the interrelationship between birth events and the migration process, basing the analysis on women's life courses and how reproductive decisions have been positioned in their movements between Morocco and Italy. I explore the ways these women, along with their husbands, construct their fertility choices by considering how different births happened at different moments of the family cycle. I reconstruct the process that led these couples to achieve their ideal number of children (2–3) or even larger families (4–6). I examine how these babies come into the world, considering pregnancy as an experience that may intersect with planning, desire, surprise, and disappointment. I also explore the employment of contraceptives and abortion. The investigation traces the context surrounding these births, highlighting how children come about to establish and ensure the household despite elements of adversity. Here, I develop the core argument of the book regarding the *transnational making of population*, showing what relational, material, and moral resources these couples rely on to be able to have and raise offspring in the context of immigration. In doing so, I highlight in the conclusion the pattern underlying the construction of high fertility in a lowest low-fertility country such as Italy.

© The Author(s), under exclusive license to Springer Nature Switzerland AG 2024
F. Decimo, *Lives in Motion*, Palgrave Studies in Mediating Kinship, Representation, and Difference,
https://doi.org/10.1007/978-3-031-65583-8_3

59

60   F. DECIMO

Keywords Migration and fertility • Birth events • Contraception •
Family planning • High fertility

The vicissitudes of the Moroccan couples I am featuring through this analysis are embedded in a sequence of family and reproductive events that develop far beyond marriage.[1] After marrying, they pursue a further key commitment concerning the birth of children and the chance to raise them in Italy, that is, to keep their offspring close throughout migration. This is not a possibility granted in migration, as an exhaustive literature on transnational parenthood has widely highlighted (Carling et al., 2012; Hondagneu-Sotelo & Avila, 1997; Mazzucato & Schans, 2011; Mazzucato et al., 2017; Parreñas, 2005, 2015). The chance to establish a family life in migration, and even more so to raise their own children under the same roof, is indeed negatively affected by several orders of constraint. In a wide range of national contexts, restrictive entry policies, ever more rigid border-policing measures, and economic obstacles increasingly force families to separate, thereby jeopardizing migrants' opportunities to have and raise their descendants while staying together (Abrego, 2014; Boehm, 2011, 2017; Coe, 2016; Dreby, 2010; Feldman-Savelsberg, 2016; Menjívar, 2006; Menjívar & Abrego, 2012). These are the circumstances underlying the worldwide formation of transnational families, that is, households whose subsistence is based on the availability of one or more family members to migrate and send monetary remittances home. These individuals relocate abroad to assume productive roles, indefinitely postponing the chance to reunify with their own family. Such a system implies the transformation if not weakening of familial belonging, particularly in terms of parenthood. This is typically the case of the migrant women called on *en masse* to provide a wide range of personal care services in a way that prevents them from keeping their own children with them (Constable, 2020). The fact that so many women are obliged to postpone pregnancy or play the role of transnational mothers under such circumstances has led scholars to hypothesize that migration has a "disruptive effect" on reproduction and intergenerational relationships, demographically quantifiable as reduced fertility (Kulu et al., 2019) and sociologically traceable to care drain and parent-children splits (Parreñas, 2015).

---

[1] This chapter expands an analysis previously published in Decimo (2021).

At the same time, however, some life-course developments seem to strengthen migrants' intentions to establish a family life based on proximity, and particularly to have children and keep them physically close in migration. In this vein, demographic research suggests that reproductive choices can intersect with mobility trajectories in unexpected ways and counter to the outcomes predicted by disruption theory (Andersson., 2004; Bledsoe et al., 2007; Kulu et al., 2019). Specifically, several studies find that reproductive choices can also be supported and facilitated by migration rather than hindered or delayed. According to the "interrelation of events" theory, indeed, movement, the formation of couples, and childbearing constitute interdependent occurrences in individuals' life courses. Various migratory flows, mainly in Europe, offer evidence to support this view, showing that the probability of childbearing may actually be higher after migration (Andersson., 2004; Kulu, 2005; Lindstrom & Saucedo, 2007; Milewski, 2007). Moreover, Andersson (2004, p. 767) finds that "international migration speeds up the continued childbearing of two-child mothers at short birth intervals and also stimulates a renewed propensity for higher-order childbearing at the longer intervals".

Such a heterogeneous range of reproductive behaviors raises several research questions that are commonly overlooked by demographic research, exploring issues such as: under what conditions does mobility restrict or foster migrants' chances to give birth to children? What possibilities do they have to live in proximity with them? Or, in other words, how are they enabled to raise their offspring in person rather than remotely?

Moroccan women in Italy represent a significant case study in this arena: not only do they display a marked tendency to entwine migration and marriage, as considered in the previous chapter, but they also frequently bear children after migration. As a result, mobility and reproduction effectively constitute interrelated events in their life courses (Mussino et al., 2015; Mussino & Strozza, 2012). In the following pages, I explore first how Moroccan women, together with their partners, have been able to overcome the detrimental effects of displacement, set up households in Italy, and have offspring. I therefore consider the process that led them to achieve their ideal family size, composed of the parents and two to three children. Second, I focus the analysis on higher fertility by also reconstructing the stories of a minority of these couples who went on to have four to six children: to investigate this phenomenon, I delve into the array of practices and values through which these women detach themselves from the fertility norm for Italy, a rate that is dramatically low (1.2 children per woman in 2022).

The analysis is based on women's life courses and the way their reproductive decisions have been positioned in their movements between Morocco and Italy. I explore the construction of these fertility choices by considering how different births took place in different moments of the individual life course and family cycle. I develop this research perspective following the arguments posed by Boltanski (2013) in his sociology of abortion and generation according to which not all "children through flesh" are destined to become "children through speech": for this social conception to take place, a process of designation must also unfold that provides for the singularization of human beings, that is, which "bring[s] new human beings into the social world" (Boltanski, 2013, p. 29). Such a process involves fecundation, pregnancy, and birth, but it is gestation that represents the crucial stage, the actual threshold through which life or abortion is at stake. Adopting this view, I examine how these children come into being, considering how pregnancies have been experienced at the intersection of planning, desire, surprise, disappointment, and refusal.

The main argument driving the following analysis is that the interweaving of migration and fertility these women deploy between Morocco and Italy stems from a transnational mobilization of reproductive resources, namely an array of relations, practices, and values that make marriage, childbearing, and the raising of offspring sustainable and attractive in migration. What I aim to make visible through my investigation is that these couples give birth to babies who are *children of migration*, that is, offspring engendered through the whole of the experiences, habitus, family histories, and interpersonal networks they have cumulated across multiple and stratified movements between Italy and Morocco. As mentioned above regarding marriages, in this case as well such a *transnational making of population* is the consequence of individual and collective efforts to maintain and develop a sphere of familial care and generation across migration. In this framing, the reconstruction of life courses I provide here is intended to shed light on the practices and values that confer relevance on the reproductive domain within the migratory process. In so doing, my investigation also aims to offer a close-grained research perspective on the way migration and fertility interrelate, one that reveals how family, kinship, and generation are so differently entangled across the Mediterranean that they could be said to underpin the perpetuation of population and, at the same time, the migratory process itself.

Living at an Intense Pace: The Interrelation of Migration, Marriage, and Births

When family matters begin to unfold, events do not occur one at a time. A trait recurrently featured in the stories I collected among the Moroccan women I met is the tight interweaving of mobility with the evolution of the family cycle: as a result, events such as marriage, migration, and child-bearing occur within a very short time frame in these female life courses, performing a sequence of rapid and interconnected steps that was repeatedly described in the accounts I collected.

Specifically, this set of Moroccan women's life stories reveals significant issues in terms of the construction of their fertility and the way childbearing occurs during different biographical phases. As suggested by the interrelation hypothesis, migration and fertility are entwined events: for 23 of the 32 women who immigrated as wives, pregnancy, birth, and movement took place within two years of each other or less. The majority of these women had just married, and so they had experienced not only fertility and mobility but also marriage in the same short span of time. Furthermore, this period also includes the time spent dealing with the bureaucratic procedures necessary to obtain spousal family reunification. At the same time, for those who immigrated autonomously or as children together with their parents, the interrelation of events involves marriage and pregnancy. Considering the sample as a whole, only one woman out of four experienced fertility as not interrelated to one or both of the other events (Fig. 3.1). After all, as one informant who participated in several stages of the fieldwork explained: "no one plans how many children they will have, [but] after the wedding everyone is expecting to have children".

Even taking into account that birth is expected to follow marriage, however, it is worth noting that these stories are affected by the migratory condition and that mobility could have a detrimental effect on fertility, as asserted by disruption theory. There is still an analytical need, therefore, to disentangle the intertwining of fertility and migration.

Indeed, the women interviewed did not romanticize the beginning of their immigration to Italy at all, instead depicting this period as very stressful, frustrating, and humiliating. The following accounts eloquently address the difficulties involved in the initial attempt to settle down and how this is often associated with the first pregnancy:

64 F. DECIMO

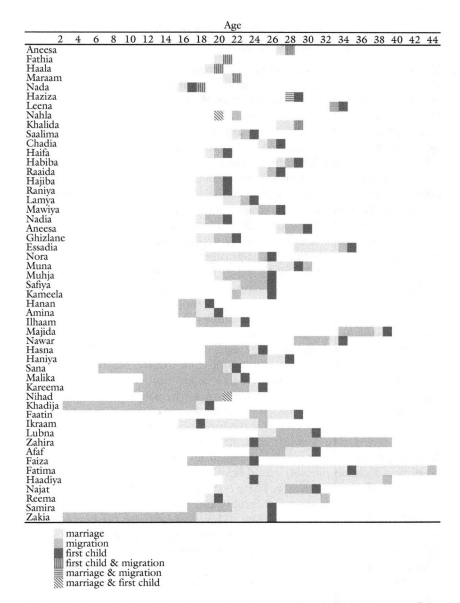

Fig. 3.1 Interrelation of marriage, migration, and first childbirth by age of the interviewed Moroccan women

3 CHILDREN OF MIGRATION: THE TRANSNATIONAL MAKING... 65

I slept next to his mother in the in-law's house! I did not have a room with a door, belongings of my own, my clothes were always in a suitcase. All the work fell to me, to make bread for 10 people! (Nora)

Our first child was born in '94, a boy. I wanted it, I wanted it ... because as soon as I arrived here, [the impact] was a real trauma! [In Morocco] I lived in a big, beautiful home with my family ... as soon as I arrived here, alone, there were not as many people as there are now, I felt bad, I could not wait to go back. (Raniya)

[The children] came right away, true. The first one, I can't tell you how hard it was. I was still young ... there were lots of things I didn't understand, in order to have a child ... I went through some difficult months, I was alone. (Hajiba)

How did you feel after the birth of your first child?

Miserable, yes, yes, I cried, with no mother, still young, 22 years old, alone in the mountains, cold. I had no one, not even neighbours. (Malika)

I came right away with him! I was still a little girl, I was not a young woman! I didn't know how to cook, I didn't know ... I learned so many things here! With my friends, with life experience. I had my 20th birthday here! But going from school to home [the transition from student to wife] [was] a bit ... demanding [...] The house was in bad shape, very cold, my husband got arthritis there, even I had back pain ... there was no washing machine, so I washed by hand and when I did laundry, the clothes ... once my hands ended up paralyzed, frozen, I went home and cried, I stood there holding my hands in front of the stove for almost 10 minutes to recover. [...] There were three gas heaters in the house but they were not able to heat it ... there was water [moisture] seeping out of the walls! We suffered so much! Me, without children, but other families had children as well ... thank God that time is over, really! (Muhja)

In these circumstances, as Raniya explained, couples may have children in order to counteract the stressful impact of their desolate settlement in Italy. However, births also come about to establish and ensure the household despite elements of adversity. On the one hand, these couples are able to surf waves of uncertainty to achieve their desired families. They are aware that precariousness is a distinctive feature of the migrant condition, but they do not perceive it as so overwhelming as to lead them to give up on having children altogether. On the other hand, childbearing offers some opportunities to access better living conditions in publicly subsidized housing, since families with children have better chances of obtaining such accommodations. In these circumstances, births serve to enhance the household rather than the other way around. This behavior is consistent with the principle, reiterated many times throughout my fieldwork, that having a family is a human right in that it represents an essential and irreplaceable source of individual well-being that meets universal human needs.

Expected children are thus the ones the couples plan to have in migration, despite all the attendant difficulties, in order to achieve their wished-for family which, as reported above, is ideally composed of parents with two or three children. In fact, the majority of the couples participating in this research succeeded in having this desirable number of children. At the same time, as the data makes clear, it is not unusual for such a number to be exceeded or, in some cases, doubled. While the first birth of a child is interrelated with the marriage and/or migratory course as an expected outcome, therefore, further births instead bring into play other tangled reproductive dynamics. When is it time for a second baby? And a third one? Which couples can afford to outstrip the number of children commonly understood to be standard?

The matter of bringing children into the world entails a multitude of dimensions, from issues of family planning to gender and intimate relations or bodily and embodied politics, as well as household socio-economic status, genealogy, and intergenerational transmission. The way emotions, desires, personal choices, and corporeal momenta take the form of conception, pregnancy, and childbirth is unpredictable. However, although each birth is engendered as a unique occurrence, through its own genealogy and singularization (Boltanski, 2013), a thread connects the reproductive behaviors of the different women, couples, and family stories comprising my ethnographic sample. Following this thread, my analysis continues by taking as a departure point the circumstances that lead to unexpected pregnancies and delving into the troubles, evaluations, and feelings that women process in considering whether or not to carry a pregnancy forward. As I aim to argue, even births that seem engendered by accident play a substantial part in shaping these fertility patterns.

Expected Children and "Gifts of Allah": (Un)Planned Pregnancies

According to Boltanski (2013), fertility is a social construction made by the government of bodies and their biology that takes place through a process and involves different phases. Not only is fecundation not guaranteed, but childbearing in particular represents an uncertain transition, and in order for a pregnancy to preclude a birth, a process of social designation must be implemented. In this vein, while among the couples I met, the first child was expected and welcomed as a seal of the marriage, as well as a development allowing the household to take root in the circumstances of adversity entailed by migration, further babies were engendered following different logics of desires and possibilities. The clues and signs of such logics can be retraced if we begin by looking at the approach to family planning the couples undertook and the order of births they pursued. Adopting this perspective, the analysis focuses on the level of the intentions, purposes, and reported behaviors. Contraceptive measures play a crucial role in this arena and are used in various ways and with different degrees of effectiveness. In a few cases, the couples described having adopted the coitus interruptus or Ogino-Knaus methods. However, the coil, birth control pill or other hormonal systems were the most frequent contraceptive devices the women involved in this research used to plan whether and when to open a window for conceiving another baby after the first one.

Female will in this choice is decisive, as Hanan—whose story introduced in the previous chapter—explains. After she had her first baby, she obtained a subcutaneous hormonal implant with the intention of avoiding a further pregnancy for at least three years. In fact, she now prioritizes earning money so as to avoid impoverishing her family and to be able to send monetary remittances to her sick mother, as the vignette opening the previous chapter already made explicit:

> Better work, money, help. Now that my mom is sick with diabetes, [to help] with insulin … She has so many health problems … That's why I don't want [more] children. Diapers, medicine—there are too many expenses. Even my mom says no, don't have children, don't have children now. (Hanan)

In general, family planning represents a widely discussed issue among couples as they go over their preferred contraceptive measures and the ideal number of children to pursue. The appropriate space between births is likewise a significant point of discussion: in most cases, this timing is based on the socio-economic conditions of the household, as explained by Hanan, as well as age criteria, both that of the parents and of older children.

However, a reconstruction of the way births shored up the histories of these families reveals that, beyond intentions, there is also a cluster of implicit choices and desires, but also mistakes, misunderstandings, and surprises, that played a crucial role in granting these households their current shape in terms of composition and dimensions. It is exactly by disentangling this web of practices, tacit preferences, and emotions that we are able to understand how the reproductive cycles of these women have been accomplished together with their partners. At the heart of these varied behaviors and the fertility rates, these women achieve certainly lies their young age at marriage and thus the odds that unexpected pregnancies will occur over the course of their reproductive lives. As Hajiba explained, such children—unplanned births often conceived due to contraceptive failure or human error—are a "gift from Allah".

> Alina is a 'gift from Allah', we say. I did not expect her, three were already enough! (*laughing*) (Hajiba)

It is a fact that many of the pregnancies occurred following the failure of contraceptive measures. These events happened more frequently when the methods adopted were Ogino-Knaus or coitus interruptus, but even the coil and pill turned out to be less than fully reliable. The only method offering them completely secure control over their fertility is tubal ligation, and this was indeed represented by interviewees as the best, final choice they opt for when deciding to permanently conclude their reproductive lives, as Khalida declared when retracing the story of her four pregnancies:

> Actually they all came like that!
> *Not planned?*
> No. All problems with methods of contraception ... (*laughing*). Every time I try one ... (*laughing*). I mean, come on ...
> *Counting your cycle days?*
> No, that's no good for me! Even the pill, if I forget it for a day ... I get a little girl! Now I've done the sterilization, what's it called ... [tubal ligation]
> *In short, you decided not to have any more?*
> Yes, yes, that's enough! (*laughing*). We got married a little late, so if maybe I say I want a child then I'll wait 4, 5 years, then have another one ...
> *How old were you when you got married?*
> 24 years old, we made promise, but then we still waited for marriage and then reunification
> *But that wasn't so old ...*
> Well, it's old for having children.
> *Did you space out the births from each other in some way?*
> Not so much, every year we have a new arrival: one more member of the family! (*laughing*) (Khalida)

A deeper analysis of the relationship these women have with contraception and its ambivalent management unveils the way pregnancies and births

3 CHILDREN OF MIGRATION: THE TRANSNATIONAL MAKING... 69

represent a hidden dimension of choices and desires with whom they maintain a silent connection. Or, to state it another way, there is a logic or implicit foundation of reason even in the contraceptive mistakes and the ways these are managed, up to the decision of whether or not to carry on with a pregnancy. It is exactly by exploring such unplanned occurrences that we gain a clearer overall picture of the fertility rates these women achieve across their life courses.

Through the account by Malika, we are able to follow a narrative thread in which coincidence and intention, desire and fate, as well as misunderstanding and awareness are unquestionably intertwined. Malika explains that she conceived her last child, their fourth one, when she was taking antibiotics that reduced the coverage of the birth control pill. What is worth stressing in this case is that she does not describe this coincidence as erroneous or fortuitous, exactly. On the contrary, there is a reference in her account to the desire for motherhood, sparked by a comparison with her female neighbors:

> Safia, I had her because with the neighbours came a desire for children (*laughing*) ... because I had said enough after Samira, but then living here with our neighbours, all with their children, saying "do it, do it!" and then the craving came over me! Safia came with the craving! The first ones, no!
> Now I don't listen to Lamya anymore! [her neighbour who has six children] ... The first ones came like that, without me thinking about it, [but] with Safia I thought about it.
> *But you have never used the pill or birth control?*
> No, for the first ones no, now yes and after Samira yes ... the pill ... but with antibiotics! (*laughing*) Because you if you take antibiotics when you don't know [without knowing that] the pill doesn't work anymore, it [the pill] doesn't do anything ... Because I didn't know that, before ...
> *And one of your children was born that way?*
> Yes, Safia! Because I always get tonsilitis ... Yes, because first I thought about it, but then she came ... now you understand everything!? the antibiotics ... (*smile of understanding*) Yes, because first I thought about it, but I did not think that was the right time ... you know, even now, what God wants to do He does, eh!!! You take the pill, you take everything, but if it [a child] is supposed to come ... (Malika)

Malika thus did not use any birth control for a long time, and so had her first three children without resorting to any birth planning methods, and only after they had been born did she begin to use the pill regularly. It was then for their fourth child, who ended up being Safia, that "the craving" arrived: she was the one who had to be "thought about", that is, sought after. The pill's efficacy was thus invalidated by taking antibiotics, suggesting that it may have been a not-so-accidental accident. At the same time, it is relevant to dwell on her concluding reflections, the idea that the very principle of rationalizing choices and planning for the future should be given less priority since the primacy of life, universally represented by natality, always transcends such measures. By concluding with such a reasoning, Malika notes that a divine plan works for just this purpose—although her narrative reveals substantial human desire and intentionality as well.

In a nutshell, although all of the women interviewed had been employing contraception quite regularly, on the one side, contraceptive misunderstandings and errors happened frequently; on the other side, the accounts I collected clearly expressed that an unexpected child is not an unwanted child. In cases where another birth is genuinely unwelcome, the interviewees do conceive of abortion as a legitimate, albeit anguished, means of interrupting pregnancy (Loghi et al., 2018). The following stories by Hajiba, Nora, and Haala shed light on the troubled decision-making process that an unwanted pregnancy elicits, but also on the availability of abortion as a possible choice and how the final decision to keep the baby was made:

16 years: they flew by … I remember yesterday … The first two, I thought that was enough. Then after seven years I had this dream: maybe I would have a girl. I waited a year with this thought, yes or no? Then I had a boy and there, enough, the idea was clear! That's enough. I laughed when I heard about pregnant women [who did not want to] and I thought, where are we? You don't have to get pregnant if you don't want to! But oh, it happened to me! I could not believe it! I cried so much, I said: no … four!? And he was there for me, I have to say. If not, I would have done something. Because there were so many of them, the other one was still young, I had no idea. He told me, I do not agree, not … if you want to get rid of it and something happens to you, you're on your own … so I was convinced … (*laughing*) … I had Alina! (Hajiba)

I blamed him because he hadn't been careful! (*laughing*) I didn't want any more! He told me: don't worry, we'll raise this one just like we raised the others. I said no, what I feel is different, the weight you know is always borne by the woman who gives birth, who wakes up at night. And I decided to get an abortion on my own without telling him. I made an appointment, but before I went in I changed my mind, I saw women, with children, I thought I cannot kill a living creature ... then when I went in, the doctor did an ultrasound, she told me look at the baby, it's three months old ... do you still want to do it [abort]? I said no, no, that's enough, I don't want to anymore. And then he was born!! (*happy expression*) [...] Then my husband and I went to get an ultrasound privately, paying out of pocket, instead of spending the usual 30 euros, and when my husband came out he said, it was worth the whole 120 euros!! (*laughing*) [...] That's enough children, now. I got the surgery, tubes tied. (Nora)

Did you ever consider terminating a pregnancy?

(*Smiles*) I thought about it with this one here!!! I swear, I thought it, because I work alone ... with 4 children, with my husband not working ... I just changed my work contract and in February I turned out to be pregnant: what? can you believe it? I thought, as soon as I tell my boss I'm pregnant, he'll say, Haala, no!

But you have a permanent contract, could they have fired you?

I do, but the company changed, as soon as you change, not even a month has passed and you say, "I'm pregnant!" I thought ... I shut up, [for the] first month, two months, three months, I say enough, I can't, I felt bad from inside, I could't sleep at night, I said, "I'll tell them, whatever happens, happens, I can't ...". But as soon as I explained with the new boss said, he went, "Haala, congratulations!!! I say, no ... I can get rid of it ...". He said, "No, no, no!!!" I thought he would treat me badly, but he gave a nice answer, you see ... thank God!

And what did your husband say? Did you talk about it?

My husband said: "I can't say anything, you know ... with these things I can't say, you keep it or don't keep it. If it were me, I'd say keep it!". Me too, I was afraid too, only I was thinking of the second one [ending the pregnancy], but he was afraid with God, you know? That if you get rid of children, you don't live well. And then what came out was a child like this! (*Haala, laughing, indicates her very pleasant and well-behaved son*) And finally I had a caesarean ... Now I'm done, closed.'

You got your tubes tied?

Yes, yes. (Haala)

In these as in many other stories collected, chance intervenes in the individual life course to inspire women, as well as their husbands, to actively, albeit controversially, reprogram their family plans and choices. The reference to abortion as an available option underscores the intentional, voluntary, and conscious character of the pregnancies they instead eventually carried forward. At the same time, these accounts portray the last newborn with enthusiastic tones, as the premium they are awarded for having been able to go through the sliding door between ending or maintaining a pregnancy: in this regard, their narratives take on engaging and passionate tones, representing such children as "gifts from Allah", as an embodiment and vital representation of their family life.

The Family Cycle and Offspring as a Reward

The analysis conducted so far has reconstructed how these women manage the risk of pregnancy and the underlying intimacy they maintain with the idea that an unplanned, surprise child may still appear in their lives sooner or later. This point is closely related to the fact that the spouses, and particularly the brides, marry at a relatively young age (Fig. 3.1): the longer the female span of fertile life, in fact, the greater the odds that family planning will end up being revised, updated, or transformed according to the desires, doubts, projects, and choices, as well as misunderstandings, oversights, and accidents, that life holds. This means not only that the possibility of an additional child may acquire increasing weight and appeal over the evolution of the family cycle, but also that the relationship between planned and unexpected children is not in any sense a dichotomous opposition but rather experienced as a continuum of opportunities.

At the same time, besides the logic of mistakes and tacit wishes that secretly foster an unexpected pregnancy, other circumstances must be taken into account to grasp how these women achieve higher fertility. A number of them (14), indeed, managed to have four, five, or six children, and looking at the way these births took place across their life courses (Fig. 3.2) allows us to identify further patterns of interrelation among births occurring after the second one. Frequently, the third pregnancy took place just after the second birth, and the fourth after the third; in some cases, three births followed each other in close succession. On the other hand, some of the more numerous families had given birth to their last child many years after the previous siblings.

A broader examination of the families in which these births occur reveals certain significant elements. Specifically, I found that the phase of the household cycle in which couples have these "gifts from Allah", and more generally the last of numerous offspring, is characterized by conjugal harmony and economic stability.

The way the last births are represented in these narratives offers evidence of the increased relevance that family, the home, and affective events

Age

	18	20	22	24	26	28	30	32	34	36	38	40	42
Raniya		o	o	o		o	o					o	
Lamya			o	o	o		o	o	o				
Muhja					o	o	o	o			o		
Nada	o	o		o		o	o						
Haala			o		o		o	o					o
Hajiba			o	o			o	o					
Khalida						o	o	o	o				
Malika			o		o	o		o					
Safiya			o		o	o	o						
Ikraam	o		o	o					o				
Habiba						o		o	o		o		
Muna						o	o				o	o	
Nora				o			o		o	o			
Reema		o	o		o			o					
Faatin						o	o				o		
Maraam			o		o		o						
Raaida					o			o	o				
Lubna							o	o			o		
Zahira				o	o						o		
Afaf							o	o			o		
Faiza				o	o		o						
Haadiya				o		o	o						o
Haniya						o	o		o				
Khadija		o		o	o								
Nihad		o		o	o								
Zakia					o		o	o					
Aneesa						o	o						
Saalima				o		o							
Nahla		o			o								
Essadia										o	o		
Kameela					o					o			
Amina		o		o		o							
Ilhaam				o	o								
Majida											o		o
Hasna				o		o							
Ghizlane			o			o							
Sana			o	o									
Najat								o	o				
Fathia			o										
Chadia					o								
Haifa			o										
Haziza						o							
Leena									o				
Mawiya					o								
Nadia			o										
Hanan		o											
Nawar									o				
Kareema				o									
Fatima									o				
Samira					o								

Fig. 3.2 Childbirths of the interviewed Moroccan women by age

have in the migratory stories of these women. Compared with the planned children couples seek to have in times of household uncertainty, these later births are represented as fruits of gratification stemming from familial well-being. The following stories from Haala, Raniya, and Nihad make clear what these couples pursue and consolidate through their reproductive choices, that is, the primacy of the domestic sphere, human relationships, and intergenerational wealth perpetuated by the act of bringing new life into the world.

Haala's household has been a dual-income one for most of its life: she has always worked since she first arrived in Italy in her 20s, and returned to work soon after her first child was weaned. She then spaced out her second pregnancy by five years using the coil, in order to be able to work. She was 25 when she decided together with her husband to have another baby, and the child arrived the year after. After this one, they waited four more years before having their third child: by this point in her life, as she declares, the will to prioritize their family and offspring was definitively consolidated. The last child was an unexpected one, and her husband, unemployed at the moment of the interview, has been the main parent to care for him while Haala, with her salary, meets the economic needs of the family, playing the role of breadwinner for all intents and purposes. The way she frames her story of having five children so as to assert the worth of her familial choices is emblematic:

> Now you find plenty of people who want to have children, but ... (*pauses*) I did this, I had my children, whoever comes, comes! (*laughing*) I always think about working, working, working, and then? (Haala)

With only a few words, Haala makes explicit a densely woven fabric of meaning and practices: she is a working woman who claims the right to have children, as an achievement prioritized over economic needs. She also stresses her will to pursue this goal in a context such as Italy where, even if everyone seems to be seeking more babies, the actual trend is to give up on having them.

Raniya's story discloses another perspective on high fertility choices: while describing the circumstances that led to each one of the births, her account allows us to understand how family may represent a nest, the playful intimacy of which underlies and fosters the desire for more children. Raniya eloquently explains that she had her sixth pregnancy in response to imperative requests from her other five children:

I had three boys and a girl. Then in 2004 our second daughter arrived, I wanted her because the [first] daughter was too lonely, she always felt excluded, she would cry because they played with each other. When her sister arrived, my goodness [what happiness!] … she was like a baby-sitter, a mom … And then in 2013 the sixth arrived! (*laughing*) My kids made me do it! Yes, yes! Now she's at home with her sister! Because my neighbour had a son and they would always say "mom, you too! why not!?" (Raniya, aged 41)

A similar case is that of Nihad, a mother of three children whose story illustrates how household material constraints, subjective desires, and family relationships may be processed in a way that leads to deciding to have another baby. Nihad had long been the only member of the household earning an income; now that her husband had finally found a stable job as well, she was able to plan on enlarging their family in dialog with both him and the children. She frankly expressed the desire to have a twin birth:

Are you thinking or have you thought of having a fourth one? Or …?
I'm thinking … but … I, my dream is to have twins
Really?
Honestly!
You'd like that?
Yes! Now I'm 32 years old, before I turn 35 … at 35 I want to …
Would you two be happy?
Yes, then that would be enough
And is this something you want, or does your husband want it, too?
Him, too … at first, he didn't, at first he said: no, that's enough, and that kind of thing … but lately he tells me: but we need at least another boy to keep his brother company [the older one]! […]
But would they like that [to have another sibling]?
Yes, yes, and when there is someone from the family, or friends we go to visit who has had some children, the youngest girl asks me: why don't we go and buy one at the hospital? (Nihad)

These different stories reveal that births stem from the wishes and future plans not only of the mothers or the couple but of the entire family. What emerges, in the end, is that the households composed of a number of children higher than that considered the expected standard (2–3) embrace these additional newborns, welcoming them as a confirmation of the order and worth of their affective, interpersonal realm. At the core of this approach to fertility and family numerosity is the value conferred on descendants, understood as a source of satisfaction and the prime recipients of their love, efforts, and projections for the future. The account by Mohammed is explicit in this regard: he reached his older brothers in Italy for work, after several years of higher education. His words clearly convey that the two children he has had represent his wealth and his fortune and are the sole beneficiaries of his material and moral efforts in a family project wholly devoted to building a family future in Italy. He has never been able to find permanent employment, only temporary jobs; even if he is able to grant a good and quite stable income to the household, therefore, his precarious economic position has had the effect of deterring the couple from having more children:

> Unfortunately, I arrived right at that moment [the economic crisis of 2008], but the education I had gotten is not lost, I didn't make a career of 18–20 years of study just thrown away! Because without those I would not have gotten to this!!! Coming here, in two years I came to understand Italian well ... then I got my driver's license here, I'm continuing to fit into society, it doesn't mean I came here closed-off and would like to make money and go back ... if you ask me if there is anything for me in Morocco: there is nothing! My fortune, my bank account, are these two (*pointing to the children*). Obviously, I hope times will improve, that I will have a [stable] job ... so I can bring to Italy 4–5 more [children]!!! We have time, in other words (Mohammed)

As his account suggests, high fertility is represented as a positive outcome that better-off families are able to afford. The well-being of these households is to be considered in relative terms, not only in a monetary sense, but conjugal complicity, household harmony, and the proud awareness of having been able to develop such life trajectories between Morocco and Italy represent further and significant elements of fulfillment. The couples, who are able to pursue this set of values and who manage to establish solid, affectionate, and numerous households in migration, represent the ideal evolution of the transnational mobilization of reproductive resources analyzed in the previous section.

In conclusion, the fertility behaviors that I have reconstructed through my analysis here delineate a family pattern that evolves through *consolidation and celebration*: while planned children are conceived within the consolidation phase of the migratory process, births after the third child are portrayed as an unexpected gift, akin to a celebration of the household with all its roles and rituals. Only a minority of households are able to achieve this goal: late age at marriage, economic instability or conjugal conflicts and divorce, as well as the diffusion of the nuclear family model based on two children, do affect the majority of couples' chances of enacting higher fertility. At the same time, however, these large families do represent a substantial minority and manifest a set of practices, values, and references that continue to constitute a social landmark.

CONCLUSION: MIGRATION AND GENERATION IN SPITE OF EVERYTHING

With the aim of exploring how Moroccan couples have been able to have and raise their children across migration, this chapter has focused on female life courses by retracing the circumstances surrounding their birth choices. As stressed in the introduction, the opportunity to entwine family-making and offspring with trajectories of mobility is far from granted for a great number of migrant women who are driven to give up on motherhood or leave their children behind in their home country. In this perspective, the analysis has focused on the significant interrelation of migration and fertility that Moroccan women, together with their husbands, have instead been able to weave across the Mediterranean. Such interrelation was revealed to be part of the broader evolution of the Moroccan migratory system based on the consolidation of households abroad and the relevance that births hold in this process. As reconstructed from the accounts collected here, individuals secure the conditions necessary to base their families in Italy by engaging with a transnational range of reproductive opportunities. This range comprises not only social and cultural resources enabling female mobility but also the formation of marital couples destined to settle down in Italy, as considered in the previous chapter. Most importantly, these resources also include shared experiences of parenting in migration that valorize fertility choices and being close to children. Indeed, offspring are at the heart of these migrants' efforts to settle down and form a household despite precariousness and difficulties, while also representing a sense of accomplishment at having succeeded in this attempt.

Furthermore, this chapter has focused on the circumstances that underpinned the construction of high fertility, as represented by women who had four to six children. My proposed interpretation of these reproductive choices calls for an updating of classical explanations based on the notion of culture, especially when they rely on portraits of closure or backwardness. Indeed, the transnational lives of these Moroccan families offer no evidence of entrapment, isolation, or deprivation. Most importantly, family-oriented norms and values were not represented by the women interviewed as anything remotely resembling a burden imposed by patriarchal norms or religious orthodoxy. Rather, they described how fertility choices—similarly to the marriage ones considered previously—are personally formulated as well as negotiated with their husbands while bearing in mind a wide horizon of mobility that constantly challenges any traditional life expectations.

Viewed through this lens, transnationalism represents not only the terrain through which these families move back and forth but also a weaving of social and cultural references that frame the reproductive choices they have made. Indeed, this transnational network of relationships enables a sharing of practices and spreading of meanings concerning the ideal family size, its composition, and gender and generational roles, all of which are based on an extremely wide range of patterns. As highlighted by this analysis of birth events along the course of the family cycle conducted so far, couples are able to pursue the ideal number of two or three children but also achieve even higher numbers of children given favorable circumstances. And this range of patterns is built and cultivated without any reference to an oppositional schema in which prolificity is widely related to underdevelopment while a lower number of children is understood as an indicator of modernity and advancement. This last point is particularly relevant considering the thread of anthropological research (Greenhalgh, 1995; Kertzer, 2005; Kertzer & Fricke, 1997) extensively documenting the multiple meanings of family planning. As these studies have clearly shown, the choice of one birth control method among others does not only have contraceptive significance; it is also caught up with repertoires of meaning concerning intimacy and subjectivity, self-control, respectability, sexuality, and fertility (Schneider & Schneider, 1991, 1995, 1996; Johnson-Hanks, 2002, 2008). The act of choosing a specific contraceptive and using it in a certain way does not pertain solely to the reproductive and domestic sphere; therefore, it is also imbued with a social value in terms of individuals' modernity, morality, and self-confidence. In this vein, the accounts presented so far reveal how the women I interviewed have been able to construct their fertility by drawing on various different repertoires of meaning and reproductive models—of prolificacy, the nuclear family or intentional birth limitation (Watkins, 2000)—that coexist side by side. The reproductive practices they adopt and the way they position themselves in relation to the marital sphere denote how confident they are with a lifestyle based on self-determination and awareness, represented by the widespread use of mechanical or hormonal contraceptive methods, the legitimacy of abortion, the way they trust in gynecology and institutional health facilities, and the involvement of husbands in reproductive choices. At the same time, such self-reliance in managing one's own body, sexuality, and intimate relationships is not predictive of family size; indeed, prescriptive directions about the number of children do not necessarily follow. In a nutshell, divergent patterns of household formation and size are deployed across transnationalism without being understood by these couples according to criteria of social distinction (Schneider & Schneider, 1991, 1995).

On the other hand, reproductive behaviors—and high fertility associated with poverty in particular—have historically represented a benchmark for racial stigma through which Otherness has been constructed and reiterated (Browner & Sargent, 2011, 2021; Fassin, 2011; Ginsburg & Rapp, 1991, 1995; Sargent, 2011; Sargent & Cordell, 2003). This is particularly true in political contexts where demography is understood as a flag for identity and the ethnicity of the population is perceived to be at risk (Kanaaneh, 2002; Yuval, 1997). However, the Italian context represents a contradictory reference case in this regard, given its historically low marriage and natality rates in spite of the narrative of the family and intergenerational solidarity being held up as hallmarks of national identity (Gribaldo et al., 2009; Krause, 2001; Naldini, 2004; Saraceno, 1994). The extent of this ambivalence is such that Moroccan couples look askance at the Italian way of making family and clearly do not perceive it as a model worth pursuing. Specifically, the women interviewed here, along with their husbands, explicitly shun the prevailing Italian reproductive pattern and low number of children it entails, deliberately positioning themselves in a contrary stance and reversing the customary position of who judges and what is considered valuable in terms of rationality and morality in these matters. In this sense, they intentionally avoid conforming to the Italian norm. In their opinion, children represent a source of value and a childless home a misfortune; they likewise do not view the protracted and self-determined inhibition of women's fertility as indicative of reasoning, control, or moderation. Rather, it is the mastery and discipline with which these women govern their lively domestic environments and numerous children that constitute distinguishing marks of merit—as will be discussed in more detail in the next chapter. What makes the difference, therefore, is not *how many* children these couples bring into the world but *how* they do so, namely within what frame of meanings and opportunity structure they pursue their reproductive choices (Bledsoe, 2002).

At the same time, it is worth stressing that this study focused mainly on the first generation of migrants. For the majority of these women and men, spatial mobility coincides with social mobility, and this intertwining of trajectories has buffered the numerous difficulties and impediments they faced throughout the migratory experience. In this sense, the children these couples brought into the world are children of migration, that is, of the unique and impermanent circumstances this generation of migrant parents has been able to navigate by weaving their existences between Morocco and Italy. The array of life events these couples experienced, with their related emotions and strains, satisfactions and miseries, and contradictions and celebrations, surely represents a legacy and a significant social milieu of intergenerational transmission, but it does not render foreseeable the fertility preferences and behaviors of their descendants. What is certain is that having children in migration raises questions related to opportunities for mobility and settlement in Europe as well as long-term projects and the very perpetuation of this transnational horizon of mobility, reproduction, and belonging.

REFERENCES

Abrego, L. J. (2014). *Sacrificing families. Navigating laws, labor and love across borders.* Stanford University Press.

Andersson. (2004). Childbearing after Migration. *International Migration Review, 38*(2), 747–775.

Bledsoe, C., Houle, R., & Sow, P. (2007). High fertility Gambians in low fertility Spain: The dynamics of child accumulation across transnational space. *Demographic Research, 16,* 375–412.

Bledsoe, C. H. (2002). *Contingent lives: Fertility, time, and aging in West Africa* (Vol. 2). University of Chicago Press.

Boehm, D. A. (2008). "For My Children": Constructing family and navigating the state in the U.S.-Mexico Transnation. *Anthropological Quarterly, 81*(4):777–802.

Boehm, D. A. (2011). Here/not here: Contingent citizenship and transnational Mexican children. In C. Coe, R. R. Reynolds, D. A. Boehm, J. M. Hess, & H. Rae-Espinoza (Eds.), *Everyday ruptures* (pp. 161–173). Vanderbilt University Press.

Boehm, D. A. (2017). "Separated Families: Barriers to Family Reunification After Deportation." *Journal on Migration and Human Security* 5(2): 401–416.

Boltanski, L. (2013). *The foetal condition: A sociology of engendering and abortion.* John Wiley & Sons.

Browner, C. H., & Sargent, C. F. (Eds.). (2011). *Reproduction, globalization, and the state: New theoretical and ethnographic perspectives.* Duke University Press.

Browner, C. H., & Sargent, C. F. (2021). *Reproduction and the state.* In The Routledge handbook of anthropology and reproduction (pp. 87–105). Routledge.

Carling, J., Menjívar, C., & Schmalzbauer, L. (2012). Central themes in the study of transnational parenthood. *Journal of Ethnic and Migration Studies, 38*(2), 191–217.

Coe, C. (2016). Translations in kinscripts: Child circulation among Ghanaians abroad. In J. Cole & C. Goes (Eds.), *Affective circuits* (pp. 27–53). Chicago University Press.

Constable, N. (2020). Tales of two cities: Legislating pregnancy and marriage among foreign domestic workers in Singapore and Hong Kong. *Journal of Ethnic and Migration Studies, 46*(16), 3491–3507.

Decimo, F. (2021). The transnational making of population: Migration, marriage and fertility between Morocco and Italy. *Journal of International Migration and Integration, 22*(1), 289–310.

Dreby, J. (2010). *Divided by borders: Mexican migrants and their children.* University of California Press.

Fassin, D. (2011). *The mystery child and the politics of reproduction.* in Browner, C. H., & Sargent, C. F. (Eds.) Reproduction, Globalization, and the State, Duke University Press, 239–248.

Feldman-Savelsberg, P. (2016). Forging belonging through children in the Berlin-Cameroonian diaspora. In J. Cole & C. Goes (Eds.), *Affective circuits* (pp. 54–77). Chicago University Press.

Ginsburg, F., & Rapp, R. (1991). The politics of reproduction. *Annual Review of Anthropology, 20*, 311–343.

Ginsburg, F., & Rapp, R. (Eds.). (1995). Introduction. In *Conceiving the new world order.* University of California Press.

Greenhalgh (Ed.). (1995). *Situating Fertility. Anthropology and demographic inquiry.* Cambridge University Press.

Gribaldo, A., Judd, M., & Kertzer, D. (2009). An imperfect contraceptive society: Fertility and contraception in Italy. *Population and Development Review, 35*(3), 551–584.

Hondagneu-Sotelo, P., & Avila, E. (1997). "I'M HERE, BUT I'M THERE": The meanings of Latina transnational motherhood. *Gender & Society., 11*(5), 548–571.

Johnson-Hanks, J. (2002). On the modernity of traditional contraception: Time and the social context of fertility. *Population and Development Review, 28*(2), 229–249.

Johnson-Hanks, J. (2008). Demographic transitions and modernity. *Annual Review of Anthropology, 37,* 301–315.

Kanaaneh, R. A. (2002). *Birthing the nation: Strategies of Palestinian women in Israel* (Vol. 2). University of California Press.

Kertzer, D. (2005). Anthropological demography. In D. L. Poston & M. Micklin (Eds.), *Handbook of population.* Springer.

Kertzer, D., & Fricke, T. (Eds.). (1997). *Anthropological demography: Toward a new synthesis.* University of Chicago Press.

Krause, E. L. (2001). "Empty cradles" and the quiet revolution: Demographic discourse and cultural struggles of gender, race, and class in Italy. *Cultural Anthropology, 16*(4), 576–611.

Kulu, H. (2005). Migration and fertility. *European Journal of Population, 21*(1), 51–87.

Kulu, H., Milewski, N., Hannemann, T., & Mikolai, J. (2019). A decade of life-course research on fertility of immigrants and their descendants in Europe. *Demographic Research, 40*(46), 1345–1374.

Lindstrom, D. P., & Saucedo, S. G. (2007). The interrelationship between fertility, family maintenance, and Mexico-U.S. Migration. *Demographic Research, 17*(28), 821–858.

Loghi, M., D'Errico, A., and Spinelli A., (2018). *Cosa ci raccontanto gli alti tassi di abortività delle cittadine straniere.* Neodemos, http://www.neodemos. info/articoli/alti-tassi-di-abortivita-delle-straniere/. Accessed 17 March 2024

Mazzucato, V., Dito, B. B., Grassi, M., & Vivet, J. (2017). Transnational parenting and the well-being of Angolan migrant parents in Europe. *Global Networks, 17*(1), 89–110.

Mazzucato, V., & Schans, D. (2011). Transnational families and the well-being of children: Conceptual and methodological challenges. *Journal of Marriage and the Family, 73*(4), 704.

Menjívar, C. (2006). Liminal legality: Salvadoran and Guatemalan immigrants' lives in the United States. *American Journal of Sociology, 111*(4), 999–1037.

Menjívar, C., & Abrego, L. J. (2012). 2012 legal violence: Immigration law and the lives of Central American immigrants. *American Journal of Sociology, 117*(5), 1380–1142.

Milewski, N. (2007). First child of immigrant workers and their descendants in West Germany. *Demographic Research, 17*(29), 859–896.

Mussino, E., Gabrielli, G., Paterno, A., Strozza, S., & Terzera, L. (2015). Motherhood of foreign women in lombardy. *Demographic Research, 33*(23), 653–664.

Mussino, E., & Strozza, S. (2012). The fertility of immigrants after arrival. *Demographic Research, 26*(4), 99–130.

Naldini, M. (2004). *The family in the Mediterranean welfare states.* Routledge.

Parreñas, R. S. (2005). Long distance intimacy: Class, gender and intergenerational relations between mothers and children in Filipino transnational families. *Global Networks, 5,* 317–336.

Parreñas, R. S. (2015 [2001]). *Servants of globalization: Women, migration, and domestic work.* Stanford University Press.

Saraceno, C. (1994). The ambivalent familism of the Italian welfare state. *Social Politics, 1*(1), 60–82.

Sargent, C., & Cordell, D. (2003). Polygamy, disrupted reproduction, and the state: Malian migrants in Paris, France. *Social Science & Medicine, 56*(9), 1961–1972.

Sargent, C. F. (2011). 12. problematizing polygamy, managing maternity: The intersections of global, state, and family politics in the lives of West African migrant women in France. In C. H. Browner & C. F. Sargent (Eds.), *Reproduction, globalization, and the state: New theoretical and ethnographic perspectives* (pp. 192–203). Duke University Press.

Schneider, J. C., & Schneider, P. T. (1991). Sex and respectability in an age of fertility decline. A Sicilian case study. *Social Science and Medicine, 33*(8), 885–895.

Schneider, J. C., & Schneider, P. T. (1995). Coitus interruptus and family respectability in Catholic Europe: A Sicilian case study. In F. Ginsburg & R. Rapp (Eds.), *Conceiving the new world order* (pp. 177–194). University of California Press.

Schneider, J. C. e Schneider, P. T. (1996). Festival of the Poor: Fertility Decline and the Ideology of Class in Sicily, 1860–1980. Tucson: University of Arizona Press.

Watkins, S. C. (2000). Local and foreign models of reproduction in Nyanza province, Kenya. *Population and Development Review, 26*(4), 725–759.

Yuval, D. (1997). *Gender & nation.* Sage.

CHAPTER 4

Copious Relationships: Intimacy and Belonging in Perilous Times

Abstract The aim of this chapter is to explore the emotional nuances and affective web of relationships that Moroccan migrants navigate in relation to their marital bonds, with a focus on the domestic realm and the ways love and intimacy are constructed and experienced. The point of departure for my analysis is a consideration of some cases of conflictual and unsuccessful conjugal choices, presented here to highlight how individuals underwent periods of crisis and the role played by their families of origin. I then pay particular attention to the wife-husband interrelationship, how it changes over time and the ways gender roles are transformed when it comes to caring for children. Relationships between parents and children are also explored, in terms of how offspring are raised. This chapter paints a portrait of those households, composed of parents and children, that most vividly illustrate the difficulties as well as the sense of fulfillment entailed in making a family and settling down across migration. In particular, I critically question the notion of intimacy, understood as a detraditionalized, "pure" relationship between individuals pursuing personal fulfillment. I instead argue that these couples achieve intimacy by pursuing what I define as *copious relationships* resulting from their ability to navigate among family extension and nuclearization, independence and reciprocal commitment, as well as kin expectations and self-determination.

© The Author(s), under exclusive license to Springer Nature Switzerland AG 2024
F. Decimo, *Lives in Motion*, Palgrave Studies in Mediating Kinship, Representation, and Difference,
https://doi.org/10.1007/978-3-031-65583-8_4

85

Keywords Personal relationships • Family conflict • Divorce • Intimacy • Household daily life

The rhythm and intense succession of life events that underpin the evolution of Moroccan migrants' trajectories in Italy are inextricably connected with the features of the personal relationships surrounding them.[1] The way these migrants conceive, nurture, and formulate their own ideas of personhood and intimacy plays a crucial role in the evolution of their existence between places and generations and in everyday life: indeed, from the couple's matchmaking to spousal reunification and family-making, this process is based not only on transnationally unfolding social networks and resources; equally crucial is the interpersonal webs through which common understandings of love, belonging, and subjectivity circulate and are shaped and shared.

In this vein, the fieldwork I conducted allowed me to map a repertory of affective meanings and nuances that is firmly centered on the family realm. In other words, the individuals' relationships with their parents and sibling(s) in the initial stage of the life course, and their spouse and children in the later stage, constitute the crucial venues through which personal feelings, desires, and crises are triggered, performed, and processed. Family relationships are imbued with intense significance in terms of people's subjective horizon of reference in that they constitute both a net of solidarity and support, and a sphere of personal attachment, expression, and fulfillment. Recognizing the importance of and exploring these relationships brings to the fore the emotional and personal capital these women and men draw on, capital that supports them in establishing and cultivating households despite the multiple lines of vulnerability and discrimination that they have had to face along their paths as foreigners, Muslims, and parents of large families.

In the following pages, I delve more deeply into the way affective life is experienced by the women and men whose stories I collected, considering the marital relationship and the way spouses construct intimacy and companionate bonds (González-López, 2019; Hirsch, 2003). Counter to a large body of literature that considers the spousal union as apart and

---

[1] This chapter expands an analysis previously published in Decimo (2022).

independent from, if not opposed to, other relational commitments, I conduct my argument here by reconstructing how these couples develop closeness and fondness while also fostering profound ties with ascendent and descendent relatives. From this perspective, I question the well-known notion of "pure relationship" developed by Giddens (1991, 1992) in which intimacy is cast as a sphere of mutual disclosure that responds only to reflexive selves, destined to last until expressive wishes have been fulfilled and not weighed down by any further relationships or responsibilities. Such an argument is consistent with the broader theorization proposed by scholars of late/post-modernity, according to which the self and personal relationships in Western societies are undergoing significant transformations in the sense of being progressively detached from consuetudinary spheres of belonging (Beck, 1992; Giddens, 1991, 1992). In contrast, in this chapter, I focus on several passages from the life courses I collected that highlight how subjects pursue their self-realization and pleasure, as well as come to terms with personal crises and interpersonal conflicts, while keeping themselves embedded in a dense fabric of personal relationships. My argument is that the couples I met achieve intimacy by pursuing *copious relationships* resulting from their ability to navigate among family extension and nuclearization, independence and reciprocal commitment, as well as kin expectations and self-determination.

Proceeding with this reasoning, I also aim to contribute to the debate on transnational marriages, which, as outlined in the second chapter, is focused mainly on the difference between arranged marriages and marriages by choice and the chances for agency and self-determination regarding spousal choice. This debate, focused as it is on the marriage-partner decision, tends to overlook what happens after the wedding, that is, how love and intimacy are experienced among these couples and the extent to which their marital life is affected by kinship involvement. I lead my investigation further into the question of conjugal choice, conducting an analysis that revolves around couples with children and an established marital history and considering how personal relationships are constructed in the long run and across diverse, embedded affective circuits. In this vein, my analysis looks at pair intimacy and the construction of a companionate marital bond as it takes place beside the bedroom (González-López, 2019) and not as a dual dynamic (Hirsch, 2003) but rather as a process which unfolds by keeping the couple anchored to other, significant personal relationships and spheres of belonging.

The chapter is structured into two main sections: I first look at cases of conflict related to the choice of marriage partner. Here, I consider how individuals deal with disappointment and rupture related to spousal choice and the role played by the family of origin in the resolution of difficult periods of their lives. The focus of my analysis is then shifted forward to a later phase of the life course taking place long after the wedding day, exploring what these couples go on to experience beyond the front door of the house and over the course of their married life. In the end, based on these empirical findings, I conclude the chapter by revisiting and dialoguing with the debate on displacement, intimacy, and social change.

Subjectivity, Conflict, and Family Adjustment

Throughout the analysis presented in the previous chapters, I focused on cases of conjugal matches and family-making based on agreement between the parties involved. I adopted this perspective with the aim of mapping the strong, multiple ties that, spanning transnationally, have sustained the formation of a marriage arena and nourished Moroccan household settlements in Italy. I now shift the viewpoint to the opposite circumstance, that is, cases of disagreement and conflict due to a mistaken choice of marriage partner, conjugal failure, and family rupture. Specifically, I consider how individuals underwent experiences of risk, loss, and the weakening of their relational references. I delve into these study cases not only to stress the fact that transnational marriages—whether arranged or by choice—might turn out to be disappointing, conflictual, or abusive just like any other kind of conjugal union. What is relevant to consider in these stories is how individuals deal with personal crisis and the way family members take on a role and position themselves on the matter. At the same time, this lens allows us to see how family norms and roles are performed and interpreted, as well as the related process of adjustment and social transformation in progress. Some cases of independent spousal choice, running counter to mainstream expectations, are particularly telling in relation to this conjuncture. By focusing on four of these, I seek to further clarify how the social regulation of transnational marriages described in the first chapter operates and how diversity, conflict, and potential deviance may be normalized and incorporated by these flexible family and kinship norms.

Mohammed's case shows that it is possible to make an independent spousal choice, without any familial approval. At the same time, his narrative repeatedly points to the relevance that this approval holds in the decision process. The story of his engagement dates to an arduous period in his life, following an accident he suffered during a summer holiday in Morocco that prevented him from returning to Italy for quite some time. In these difficult circumstances, he met Aneesa, a woman who was particularly supportive of and tender toward him. They fell in love and decided to marry. As is customary, the couple sought their parents' approval to follow through with their plans. However, due to some negative gossip that was circulating about Aneesa's family, Mohammed's relatives refused to give their approval:

> Because for us, for tradition, the first thing is who you have to talk to, and that is your father and mother. He said: "yes, I know that you are the last one not yet married, if you want, I'll look for [a wife] for you", and so on, because he already knew who she was and wanted me to change my mind. But I said: "okay, dad, listen, I didn't call you to ask you to find me a woman, I have a woman. Look, I'm still in love with her, I'm going to marry so-and-so, I'm fine with her, I liked her from that moment that I was having that crisis" and he told me: "okay, talk to your brothers and sisters, if they agree, I agree too". They did this talking to each other but they all expressed a big "no", they didn't want it, none of them … So, I started calling them, one after the other, and asking: "what do you think?" and everyone "no, I'm sorry, maybe you can find another woman. […] I went to my dad's again and I brought him two people he respects, in short, people he can't say no to, and he always said the same thing: "I'm not saying he can or can't marry, he lives his own life, but I don't approve". And at that point I really started to get angry. […]
>
> We had a small wedding, I was really proud to do it! I thank God that they all didn't agree, so we could have a wedding like that, where we didn't spend thousands of euros that then you have to go back and pay in instalments! And then everyone changed their minds, they all approved! […] A great point of pride for me is that really, I have nothing, I have only these three people [wife and two children], but everyone really respects me now. (Mohammed)

In the end, Mohammed was able to bring his family around to a positive view of the union, as described in the final part of his story. The relational struggle he waged to achieve this result and the words he uses to describe his satisfaction clearly convey the social value of the marriage, as symbolic capital and a crucial node in his actual network of belonging.

A different perspective is offered by Nawar and Kameela, two women who both found a new partner after a previous experience of conjugal failure and divorce. Nawar's case is the potentially more deviant one: her first marriage was to an Italian man, conducted against the will of her parents. The relationship did not work out, so she decided to divorce him. Sometime later, while in Italy, she met a man who was from the same village as her. They began to date and, when they decided to become engaged, they announced their decision to their respective families by phone. The families gave their approval for the betrothal at a distance, celebrating it publicly in the village in the couples' absence. An actual wedding did not follow because Nawar became pregnant, and caring for the baby absorbed most of the couple's attention. The day the families in Morocco publicly declared the formation of the couple, Nawar and her groom were in Italy: no ceremonial or ritual marked that day for them; she simply moved her usual ring to a different finger, as she ironically stresses. Being married or not does not make a difference in their case, as she asserts. The ritual held to legitimate this union evidently concerned the kin and local community more than the individuals involved. Social norms are flexible enough to allow her to introduce herself after the fact, just "because people had to know who the lady is", while another reunion took place in the Italian town where several kin from the groom's side were based, as she recounts:

> We were in Italy and his family went to my parents' to get the consent. Then they held the engagement without us. We didn't go, we didn't have time. The official marriage is another thing, there you read from the Koran between two [Fetiha] and my father gave consent to his father in front of all the men. But I didn't want to go through with the marriage act ... I wanted to do everything slowly, but then the child arrived! On the day of the wedding, we exchange rings just like you do here ... but I have not exchanged any rings! I put the ring here and took it over here [moving an old ring from one finger to another].
>
> Then I went [to Morocco] because people had to know who the lady is, why not let us see her, she's a bit ugly. Then afterwards we did one in Perugia, too, so this way if I go there, no one will gossip. He wanted to do it, he invited all his family, because he has a large family. We did it and everyone knows that we are husband and wife even if we don't have a marriage certificate. (Nawar)

More conventional but analogous is the story of Kaamela. She immigrated to Italy with her family. When she was 23, on the occasion of her sister's wedding held in Morocco, she was introduced to her cousin (the son of a paternal aunt), a man put forward and supported by her father as a possible husband. Kaamela agreed, and the kin arranged the wedding to be celebrated the year after, again in Morocco. Once in Italy, Kaamela's new husband convinced her to quit her job and move to another town. Now living at a distance from her relatives and unemployed, Kaamela became pregnant and, at the same time, realized that the union she was in had no potential for love or trust. She thus decided to leave him and move back to Trentino:

> *Did your family support you?*
> Yes! My dad actually didn't [even] want me to quit my job, he said "stay at your job".
> *And when you wanted to divorce, your family was on your side?*
> Yes! yes! There were also some serious problems and so, even though I was pregnant, I had to leave it anyway. Better! After not even a year I returned to Trentino. He was never around, he wasn't there at night, he wasn't there during the day! How could I go on with this man? I don't even know where he works, who he spends time with, I don't know what he does! (Kaamela)

She returned to live with her parents and siblings for several years until she was able to achieve independence and move with her son into a new apartment. Kaamela raised the baby by herself, as the man had disappeared from their life. When the child was seven years old, she asked him if he would like to see her married and to share their home with a "new" father. He enthusiastically agreed, and so Kaamela, now 35, decided to look for a new opportunity to marry:

> *How did you two meet?*
> Once again in Morocco, we met on our own! That was last time [an arranged marriage], this time, no thanks! We met in his city, because my mother is from his city. (Kaamela)

When I met Kaamela, she had been married for approximately three years and had given birth to a second child in the meantime. I interviewed her at their house, in the presence of her husband, a dedicated and kind man who took care of the baby so we could have our meeting. He barely spoke Italian, and when I asked to Kaamela if he was able to find a job she replied that it was better he stays home to take care of the children since her working hours were quite inflexible.

Furthermore, Kaamela story reveals an important dimension of social change in terms of family structure and the line of descent. Specifically, the role her family of origin played in protecting and supporting her decision to divorce and backing her as she moved toward progressively greater independence points to a shift toward matrilineality. It is a fact that her son and the new family she is going to form are destined to orbit around the mother's kinship group rather than the father's. An even more explicit story in terms of distancing the kinship lineage away from patrilineality is Kareema's account. She also immigrated to Italy as a child, together with her family and siblings. When she was 24, she married a cousin on her mother's side, in Morocco, and then settled down with him in Italy. While she was pregnant, she came to understand that he planned not only to bring his parents to live with them in Italy but also to follow patrilocal norms and, according to Moroccan laws of succession, considered the daughter Kareema was about to bear as not fully entitled to a future inheritance by virtue of her sex. Kareema decided to divorce him and was supported by her parents so as to keep the baby girl with her. To this end, it was essential that the birth not take place in Morocco, where family law grants the father more rights over children. The account of her daring journey to return to Italy is unquestionably compelling:

> My first husband, Nejwa's [her daughter's] father, is my cousin on my mother's side. My mother's side [of the family] never get along, they have strong personalities … we were engaged for 3 years, I took him to Italy, met him there, he was very good. One year older [than me]. Then he changed. Because he is someone who listens to his family. I got pregnant and when I was 3 months along, we broke up
>
> *What were his parents telling him?*
>
> That he had to work and send money to them. It was bothering me, I was working 8, 9 hours a day, we were living with my parents, my father and mother, they had given us a room … and instead his plan was that his relatives would come up, we would get a house together and it should not be in my name … in the case of separation, my daughter would not get a whole part, she would get only half, things like that! I wanted to argue, but he was convinced that only half should pass to his daughter. It is the law. And they also say it's that way for religious reasons, but I don't know if that's true. It is not like it is in Italy, where men are like women. I said enough, that's

enough, I can't take it anymore, he took his stuff and went to his sister's, but he didn't want to grant me the divorce! So I went to Morocco, to a judge who knows my father, and the judge advised me to wait until the baby was born. But my daughter was born neither in Italy nor in Morocco, but in Spain. It was a mess! Since for maternity leave I get two years, almost, I went to my mother's in the summer and my daughter was supposed to be born at the end of October in Italy … But while I was there the pains came, my mother called the doctor at home and she says I'm giving birth! So, I think, if her father finds out and comes here, he won't agree to sign … I couldn't take the plane because I was nine months along and they wouldn't let me get on, so my father suggested we leave by car. I arrived at the border, and then got on the ship, in labour, in pain … and then we ran to the first hospital … I was lucky! I gave birth in a place called Frontera del Nacimiento,[2] right at the edge of Spain! (Kareema)

In conclusion, these different stories make clear that parental and relatives' involvement in the marital life of individuals is not limited to spousal choice, since their engagement could be required after the wedding as well—especially in the most deviant or conflicting of relationships. This fact suggests a sliding equilibrium between subjects' desires, belonging, and social obligations. Kinship norms, roles, and expectations do constrain individual will to some extent, but they do not prevail in all cases or circumstances: as the stories of Mohammed, Nawar, Kaamela, and Kareema have shown, relationships with either the family of origin or the partner, or both, can be cut off completely if they prove to be a source of individual unhappiness and negative social capital. At the same time, the reconstruction of these four trajectories indicates that individuals achieve personal, intimate satisfaction by reconnecting their own choices to the family and kinship network of belonging rather than by severing this connection. Mohammed does not hide his pride in recounting, at the end of his interview extract, how his relatives changed their opinion about his wife and in citing the social recognition the couple now enjoy. Nawar left her first Italian husband—a man who, by the way, her parents disapproved of—and became engaged to a man from the same small Moroccan village as herself, involving parents and relatives in this choice and revealing the flexibility of "traditional" marriage norms. Kaamela was negatively affected by her father's decision regarding her first, arranged marriage, but the same

---

[2] The original name of the place has been modified, leaving its original evocative meaning intact.

94   F. DECIMO

father—together with her mother and sibling—materially supported her resolution to divorce and pursue a new, independent life by herself. Similarly, Kareema was encouraged by her parents to distance the man she married from her and her daughter's lives. All of these interviewees have been able to merge the critical conjuncture between self-determination and social rules, negotiating traditions as well as navigating through complex, multi-faceted, and articulated family and kin relationships rather than dismissing or circumventing them. On the other hand, the magnitude of these processes is so great as to drive profound changes in family structure and interpersonal relationships, making gender and intergenerational bonds closer and more tender, as revealed by the possibility to shift toward matrilinear descent for daughters.

ORCHESTRATED HOUSEHOLDS: INTIMACY AND FAMILY LIFE ACROSS MIGRATION

With the stories of Kaamela and Kareema, this analysis recognizes the conflictual dynamics that may fray a marriage to the point of destroying it. At the same time, it was Kaamela's interview in particular that opened a window onto a sphere of domestic warmth, allowing us to glimpse how a good marital relationship is expected to work. Adopting this perspective with the aim of understanding how intimacy and home are constructed across migration, I have explored the repertoire of affective and relational modes, practices, and meanings that these women and men deploy in anchoring their conjugal life in Italy.

To grasp the way closeness is experienced among the couples involved in the research, it is important to consider that, as most research participants reported, a greater degree of mutual understanding and love is expected to develop after marriage, not before. Two further features repeatedly appear among these families once settled in Italy. First, with the exception of a very few households that included the husband's elderly and unwell mother, the couples involved in the research avoided the norm of patrilocality (Lievens, 1999) to instead establish a nuclear family. Displacement and distance from kin are obvious constraints on the formation of extended and multiple households, but the need for private space has acquired value in and of itself in the shape family settlement has taken over the course of Moroccan migration. Second, as I highlighted in the previous chapter, these couples expect to have children, and migration is

4 COPIOUS RELATIONSHIPS: INTIMACY AND BELONGING IN PERILOUS... 95

perceived not as a family cycle event that might jeopardize this aim but rather the contrary. Given this expectation, from the very beginning of their marital story, couples include children in the way they represent and put into practice their sphere of intimacy. The formation of such nuclear households with children, in many cases also numerous children, brings the spotlight to bear on the couple and its role as the core of familial organization. Indeed, when elicited to reconstruct the way their daily life is organized, all of the interviewees described their husband-wife and parent-children relationships as acquiring new, unexpected relevance. The following accounts by Muhja, Khalida, and Malika, respectively, describe different forms of marital and paternal participation starting from the moment of birth:

> The first delivery was a little hard, because it was my first. My husband was there with me, only my husband ... with me, always [*smiling gently*], always with me in the delivery room ... (Muhja)
>
> *How would you describe the division of labour at home?*
> We do everything, he gives me a hand, too. If I'm busy he takes the kids out, when he was not working he always took them. That way he gives me some time to rest a bit and do other things at home.
> *Does he change diapers, too?*
> Yes, yes, he does everything.
> *Giving baths?*
> Yes, yes, he even likes to!
> [...]
> *Did he watch all the births? Was he there with you in the delivery room?*
> Yes, all four of them.
> *So, dad has always helped?*
> Yes, he does everything, even though that's not very common with us. Boys work and bring home money and you have to do everything at home ... my husband does not agree with that. [...] Others are changing, too. They come up with concepts, with ideas and that way, then, over time they change, when they see that both of us need to do all these things having to do with the family, they change their minds, their behaviour as well. (Khalida)
>
> *Was he there with you during the birth?*
> My husband was there, always! [...] I think it's better if the husband sees everything, huh! See how his wife is, you know! My husband was always there all the time! I did his whole hand, like this [crushing it]!
> *After the childbirth, who was there to help you?*

96   F. DECIMO

> Nobody, even my sister was sick in the hospital in that period, no one was there! My husband did all the cooking.
> *Was he able to? Did he do things?*
> Yes, yes, first with his mom, because she did not have any girls and so all the boys knew how to cook, everything! (Malika)

As these interview passages suggest, husbands participate actively in every stage of the family cycle, particularly the stages involving children. A highly significant element in this process is the variety of tasks that these men assume, from assisting their wives during childbirth to taking care of newborns. It has become very common for fathers to be present in delivery rooms, but more generally, the interviewed women offered innumerable descriptions of paternal dedication, and the men who participated in the research echoed this point. In the absence of the kind of close female figures who would traditionally help during childbirth and puerperium, these couples were driven to develop greater closeness and cooperation than they might have in Morocco, as Khalida explained. In this regard, it is interesting to see how Malika proceeds with the description of her relationship with her husband, as it sheds further light on the deep sense of emotional understanding and intimacy that can underlie these couple relationships:

> *Do you talk to him?*
> Yes, yes, he gets it … how I ever found someone who … what's the word …
> *Understands?*
> Yes, he understands me so well! Everything, even if there is something inside me that I do not want to tell anyone, he understands that it's there … and I talk a lot with him, I say everything, secrets … he does, too … because I'm not so lucky with people, because I talk, I say everything, I don't hold back, that's just how I am, I say what's inside my heart, and this always causes problems (Malika)

The key point to stress here is that the very requirements of organizing such a complex family life in displacement nourish this closeness of the nuclear family. There is no doubt that orchestrating everyday life was challenging for the interviewees, characterized as it was by intense extra-domestic work schedules, caring for children, and managing the domestic environment. Most of the families had a single breadwinner, with the women acting as housewives and viewing the home as their purview and responsibility. They put this view into practice by taking a leading role and delegating various activities to other members of the family unit, as Nada, Raniya, and Khalida describe:

> *Does he help you with the housework?*
> I personally prefer that my husband not even help me move a glass, because poor thing, he wakes up at 4:00 a.m. while I stay in bed, then comes back from work tired, because selling in the market is a bit difficult.
> *But beyond housework, if you have worries, thoughts, do you talk with him? Does he help you?*
> Yes, sure! Yes, yes, for the serious things, the big things, he's there! I do doctors, the mail, groceries, he does all the big things. (Nada)

> When he's home he helps me! Then he does things outside the house, bureaucracy … for grocery shopping I have to be there with him. At home for example he moves things … the children help. Even the boys help, they do almost more than the girls! (Raniya)

> Now that they are older, I wake up in the morning, make breakfast. They [the children] eat breakfast, clear the table and then they make their beds and tidy up as long as they can, then go to school by themselves. (Khalida)

On the other hand, in some cases, it was the women who worked. This was true of Haala, who was her family's sole breadwinner during the time of the interviews, and Nora, who described for us the complex juggling of schedules through which their week was orchestrated:

> *And does he help out with the kids?*
> Yes … It's just me doing it, just me working … Yes, yes, he is good, he loves him [pointing to their small child]. He changes him, feeds him … I leave at 8:00 in the morning! (Haala)

> *How do you organise daily life?*
> To adapt his job and my changed schedule, he works [has shifted his hours] from morning till night to be able to. At 2:00 in the morning he finishes work, showers, he arrives home at 3:00 a.m. But then he goes out again to get the kids from school, I work in the morning and finish at 4:30 pm. He starts working at 5:00 p.m. Me, from the beginning to now it's always been complicated. We have no other organization. He was glad I found work. For five years, I have been working at the same company, in the canteen. We barely cross paths, except on Saturdays, because I work six days a week, I only have one day off. You have to help the children bathe, do their homework, cook, clean up! (Nora)

98   F. DECIMO

These narratives shed light not only on interactions between husbands and wives but, more generally, on the key role that the presence of children plays in strengthening both the couple and the affective climate of the family as a whole. Children are clearly the main recipients of the entire household's financial and emotional resources as well as its sense of identity. They are subject to constant material and immaterial attention and give rise to a relational dynamic alternating between control, closeness, and intimate familiarity, often characterized by playfulness and irony. In the following account, Nada describes her strategy of maternal control over the older boys while in the background, the TV broadcasts a Moroccan video of a very famous children's song portraying a father and his little daughter having a lot of fun playing together in a shiny, modern living room:

> *What do they tell you?*
> For example, if they're late, I ask where have you gone and they tell me to the coffee shop, to the pub ... then about the girlfriends I don't know ... but I look at their chats, and one of them knows that I look at their messages and also the photos they take and so he deletes them before entering [the house] ... but the other one doesn't, he leaves everything and then he says "I know you've looked, I know you've seen everything ..." (*laughing*)
> *So, you joke about this? What's the climate like in the house?*
> They behave with me, because I'm like a friend, because they grew up with me, you know I was 17 when I had my first child ... and they talk to me not like I'm their mom, they call me by my name, only sometimes mom ... (Nada)

A similar atmosphere of humor underlies the relationship between Nihad and her children, who joke with her about another potential pregnancy:

> Since I always have this big belly ... for months and months my children have been like: "mom, [is it a] boy or a girl?" "Neither a boy nor a girl". "But, really, mom?" I say: "neither a boy nor a girl, this is just bread!" (Nihad)

Such portraits of home and family interactions clearly convey that the large size of these households reinforces the dynamics of nuclearizing domestic life, renegotiating gender roles, and valorizing interpersonal closeness, and not the other way around. At the same time, the way intimacy is shaped between partners and in parent-child bonds is quite distinct from Giddens' model of "pure relationships" (Giddens, 1991, 1992). The reconstruction of these couple' stories conducted so far—from the involvement of family and relatives in the marriage arrangements and approval to the constitution of prolific households—reveals a different ideal of intimacy that, as introduced at the beginning of this chapter, I suggest, is based on *copious relationships*: far from seeking satisfaction by enhancing the couple itself as an interpersonal sphere devoid of external bonds or constraints, the migrant couples involved in this research experience love and closeness as inextricably intertwined with familial numerousness, complexity, and extension.

These women and men are not more interested in self-expression or mutual disclosure than they are concerned about the household's day-to-day organization. Their sense of fulfillment derives from the ability to navigate through their full and lively everyday familial lives without allowing things to become too chaotic or enter into a state of crisis. According to this view, intimacy is related to the experience of belonging, of being at home and part of a family, of enjoying domestic order and warmth and the satisfaction of having been able to achieve such a goal despite displacement. On the other side, while these are the basis of the household's daily routine, family commitment and belonging are also deployed on an annual schedule by embracing the families' large transnational and translocal kinship through ritual, planned summer trips to Morocco. Samira is particularly clear in expressing this aspect when mapping her usual kin-tour from Casablanca, where they start, to southern Morocco, specifically her place of origin in the Guelmim area:

> We basically take a tour of everything! Now there is grandma [her mother's mother] staying with the aunt in Casablanca, so we have to visit her there. Then there are other uncles in Marrakesh: there are two of my mom's sisters plus three of my dad's brothers, so we do that [stopover] as well, anyway it's on the way, we go down from Casablanca, get to Marrakesh and then we go to the desert, which is where my paternal grandfather and other uncles live. *How do you get around? Do you go by car, or ...?* We go by car, or by plane and then by bus. Casablanca and Marrakesh are easy but then there's almost a day to be spent in the desert, partly because the roads are what they are. *Do you go all together with your parents?* I used to go down with my parents, but now that I'm married, I go with my husband. (Samira)

This picture assumes further clarity in light of the settlement context of these families and the intersection of the multiple, diverse difficulties they face in carrying out their daily lives in Italy. Many interviewees reported encountering a creeping, lingering Islamophobia in multiple areas of life, at work, when shopping and in interactions with their children's teachers. Some episodes were actually quite frightening: to give an idea of the social climate surrounding these family stories, Muhja and her husband, together with the other families living in the same housing complex, were victims of an attempted arson attack by a group of neo-Nazis.[3] However, other aspects also render these households frail: increasingly insecure employment conditions stemming from the 2008 economic crisis that have caused many to lose their jobs and face new, more precarious and demanding kinds of employment contracts; the legal uncertainty generated by legislation that fails to guarantee citizenship to children born in Italy and makes work a prerequisite for residence permits; and the housing issues faced by families unable to access either public housing programs or the highly segregated commercial real estate market.

It is by recognizing the enduring fragility of their life settings and their contested belonging that we can ultimately grasp how family life confers importance and stability on these women and men's migratory trajectories. From this point of view, it is clear that taking care of oneself and one's spouse, raising and educating the children, and maintaining and supplying the home space all constitute dimensions of life that expand and anchor individuals' private lives. It is from this intimate, individual center that the construction of life outside the home can be braved, organized, and directed. By granting primary importance to the family domain as a space of generation, affection, play, and caring, these families establish a social order and ethics for their migratory paths that rank other concerns as less of a priority.

---

[3] The attack was eventually claimed by the group Fronte Skinheads and covered by the main national newspapers. In order to protect the anonymity of the interviewees, I have chosen not to provide any other details.

Conclusion: Family, Displacement, and the Everyday Politics of Intimacy

Couple formation, marriage, trust, and love represent diverse dimensions of affective life, the interweaving of which, whether tight or loose, conflictual, harmonious, or inconsistent, represents an intriguing field of inquiry. Scholars have focused on this research strand in the last few decades by reviewing the processes underlying the formation of intimate relationships based on self-expression, mutual disclosure, and closeness, understood as a peculiar trait of interpersonal life in the late/post-modernity of Western societies. Specifically, through the hypothesis of detraditionalization (Beck, 1992; Giddens, 1991, 1992), it is argued that couple formation has been—or is destined to be—disengaged from the roles and commitments that customary spheres of belonging (family, kinship, community) have long placed before individual needs and desires. According to this view, the intense social changes characteristic of the contemporary moment have detached individuals' destinies from institutional spheres of identity. This shift implies not only leaving it up to individuals to decide, on their own, how they will conduct their personal lives but also prioritizing the need for self-realization over the need for care and reproduction in intimate relationships. This is the core theme underpinning the notion of "pure relationships" developed by Giddens (1991, 1992) which, as mentioned at the beginning of this chapter, is understood as the prioritization of reflexive selves and individual expressive wishes over any other kind of interpersonal responsibility and commitment.

The notion of detraditionalized intimacy has been subjected to extensive critique by various scholars (Jamieson, 1988, 1999; Cherlin, 2004; Green et al., 2016; Gross, 2005). Jamieson (1988) in particular provides an exhaustive discussion of each of the dimensions that allegedly drive the transformation of personal relationship across modernity, dismantling the unproven assumptions and selective use of references underlying such arguments. This critical view shifts the focus back to power and diversity, calling into question how inequalities play out in personal relationships. Above all, as argued by a wide range of studies, there is no evidence that "pure relationships", established to fulfill only the reciprocal needs of understanding and knowing, are actually supplanting that mix of love, care, and dependency that continuously distinguishes personal involvement in intimate bonds (Jamieson, 1988, 1999).

102   F. DECIMO

Similar criticisms also emerge when scholars scrutinize another notion related to intimacy, that of romantic love understood as relationship based on mutual attraction, likenesses, and shared emotions. Like the ideal of a pure relationship, romantic love is thought to challenge the consuetudinary rules of couple formation governed by kinship according to pre-established, convention-bound roles of social reproduction (Goode, 1959). As Illouz (1997, 2007) argues, love and its idioms ideally work under capitalism as a transgressive, liberating catalyst. In reality, however, rather than projecting the expression of the self toward an emotionally unconstrained social horizon, such sentimental surges actually embed personal life within structures of affective life based on rationality and consumption. In this framework, according to Illouz, the rituals of leisure and hedonism related to the ideals of love end up being deeply intertwined with the disciplining of production and models of self-sufficient individualism.

On closer inspection, the notions of love and intimacy based on pleasure, desire, and self-gratification not only constitute ideal, abstract patterns which are destined to collapse when brought face to face with the asperities and asymmetries couples experience at the moment that real-life, intersected inequalities come into play (Jamieson, 1988, 1999). Also and more specifically, these understandings of affective relationships are ideological in that they selectively reify certain historical processes stemming from the economic, demographic, and social transformations that took place in Western Europe and North America in the nineteenth and twentieth centuries. This critique is not tantamount to denying that such an ideology is powerful and has elicited widespread images, fancies, and behaviors (Hirsch & Wardlow, 2006; Padilla et al., 2007). The point is that the more these practices and imaginaries circulate, the more they can be seen to readapt and transfigure the original ones instead of replacing them. In so doing, they engender an ever-new repertoire of ways of experiencing and representing love and intimacy. Adopting a global perspective, it becomes clear that there are a multiplicity of meanings and means through which individuals make and unmake personal relationships, affective bonds, intimate connections, and family life. The very idea of social change, understood as linear transformation, runs the risk of misleading us: if the research aim is to depict and typify a certain trajectory of evolution, indeed, it fails to disentangle and portray the multiple and uncertain processes through which intimacy and relatedness (Carsten, 2004) are configured or dissolved.

Marriage across borders thus represents a fruitful nexus for exploration in that it brings to the fore new, different dimensions through which interpersonal intimacy is constructed in today's ambivalent times of globalization and contested mobility (Beck & Beck-Gernsheim, 2010). Transnational couples in particular simultaneously portray all the contradictions, challenges, and questions arising from the varied, entangled matters of affective life: How is intimacy achieved, experienced, and represented in displacement? How are gender and generational roles constructed and transformed in contexts of social and cultural complexity? How do class, nationality, religion, and race intersect in enhancing or inhibiting individuals' emotional and familial fulfillment?

In an effort to dissect this dense knot of issues, I have delved into the Moroccan marital stories presented so far according to two lines of inquiry. First, I considered the role family and kinship play in transnational marriages by looking beyond their function as a form of social capital that supports and influences their marital choices. Rather, I reconstructed the value that the subjects have attached to their relatives' approval in the making of the marriage decision, focusing on individuals who made independent or conflicting choices in this regard. What emerges are stories of self-determination driven by individual will and the ability to juggle articulated family and kin relationships instead of setting them aside. Indeed, all of the interviewees describe genuine involvement in their extended families as an indispensable and irreplaceable sphere of solidarity, identification, and belonging. Most importantly, such interpersonal dynamics unfold across migration, through a process of spatial distancing from the family norms governing gender and intergenerational roles in Morocco such as to elicit creative adjustments of the household arrangement and way of life. The shift toward a matrilineal and matrilocal lineage of belonging and family solidarity was thus revealed to be highly significant in terms of gender roles and female empowerment, a finding that is consistent with other studies in this area (Lievens, 1999).

Secondly, I explored how these couples have shaped their personal relationships by crossing the thresholds of their houses and documenting the way their daily family lives are orchestrated. The analysis has scrutinized how individuals represent a good relationship as such, which dimensions constitute it as appealing and satisfying, and what configuration of roles they develop to put their expectations into practice. What appears again and again is a picture of nuclear, prolific households centered on the conjugal couple and its relationship with the children. Wife-husband bonds are described as more close-knit and balanced than they might be if based in Morocco. However, the innumerable demonstrations of intimacy they display were found to be embedded more in the lively schedule of these large-sized households than in a mere couple relationship. These husbands and wives do not represent intimacy as a dimension which is detached from family bonds. Rather, they pursue what I depict as *copious relationships*, embedded in family and kinship spheres and resulting from their ability to navigate their networks of belonging and the expressive needs of individuals across the times and places of migration. The capacity to maintain this precarious balance together, in agreement, represents a source of fulfillment and gratification that underpins and consolidates these couples as intimate, close, and harmonious. In so doing, they are able to ride the waves of late/post-modernity rather than being pulled under, that is, to circumvent the intersection of multiple dimensions of disadvantage and the risks of pauperizing family life that come with processes of global mobility, widespread care drain, and the increased segmentation of reproductive opportunities.

In conclusion, understanding the way these couples conceive and experience a sphere of intimacy as a realm of *copious relationships* enriches the discussion about family-making across globalization and displacement as a critical transition. Love, care, and reproduction do indeed represent spheres of life that are at risk by virtue of their intersection with various regimes of (im)mobility and the ambivalences of globalization. Viewed in this way, marriage proves to be a process more than an institution, since the uncertainties it implies (Carsten et al., 2021) are amplified by movement and displacement: on the one side, being married does not necessarily mean staying together; on the other side, when a migrant couple comes together, all the rules and roles underpinning the consolidation of marital bonds must be reinterpreted, adjusted, and transformed according to a script that is essentially unwritten and open-ended (Maunaguru, 2019). In this scenario, the chance to establish a household with children, complete with its everyday life tasks and rituals, is certainly not taken for granted; women and men instead put a great deal of effort into achieving it, first by overcoming periods of individual precarity and loneliness and then by navigating marital formation and consolidation. Intimacy is understood in this perspective as a *political* process caught up with the chance to establish a sphere of care, belonging, and love: it entails the need to anchor and secure one's emotional life while keeping and nurturing affective ties, and as such it is metaphorically related to the search for home, understood as the person's own place of confidence and comfort. Interpreted in this way, intimacy concerns the creative, tumultuous, and absorbing process through which family as a bustling tangle is composed; this representation is quite distant from the framing of interpersonal closeness as the unconstrained fluctuating of relationships through the shimmering streams of reflexive selves.

REFERENCES

Beck, U. (1992). *Risk society*. Sage.

Beck, U., & Beck-Gernsheim, E. (2010). Passage to hope. *Journal of Family Theory & Review, 2*(4), 401–414.

Carsten, J. (2004). *After kinship*. Cambridge University Press.

Carsten, J., Chiu, H. C., Magee, S., Papadaki, E., & Reece, K. M. (2021). *Marriage in past, present and future tense*. UCL Press.

Cherlin, A. J. (2004). The deinstitutionalization of American marriage. *Journal of Marriage and Family, 66*(4), 848–861.

Decimo, F. (2022). Copious relationships: Transnational marriages and intimacy among Moroccan couples in Italy. *Journal of Family Studies, 28*(4), 1255–1271.

Giddens, A. (1991). *Modernity and self-identity*. Polity.

Giddens, A. (1992). *The transformation of intimacy*. Polity.

González-López, G. (2019). *Erotic journeys: Mexican immigrants and their sex lives*. University of California Press.

Goode, W. J. (1959). The theoretical importance of love. *American Sociological Review*, 38–47.

Green, A. I., Valleriani, J., & Barry, A. (2016). Marital monogamy as ideal and practice. *Journal of Marriage and Family, 78*, 416–430.

Gross, N. (2005). The detraditionalization of intimacy reconsidered. *Sociological Theory, 23*(3), 286–311.

Hirsch, J. (2003). *A courtship after marriage: Sexuality and love in Mexican transnational families*. University of California Press.

Hirsch, J. S., & Wardlow, H. (2006). *Modern loves: The anthropology of romantic courtship & companionate marriage*. Macmillan.

Illouz, E. (1997). *Consuming the romantic utopia: Love and the cultural contradictions of capitalism*. University of California Press.

Illouz, E. (2007). *Cold intimacies: The making of emotional capitalism*. Polity.

Jamieson, L. (1988). *Intimacy*. Polity Press.

Jamieson, L. (1999). Intimacy transformed? *Sociology, 33*(3), 477–494.

Lievens, J. (1999). Family-forming migration from Turkey and Morocco to Belgium. *International Migration Review, 33*(3), 717–744.

Maunaguru, S. (2019). *Marrying for a future: Transnational Sri Lankan Tamil marriages in the shadow of war.* University of Washington Press.

Maunaguru, S. (2021). (Un)certain futures: rhythms and assemblages of transnational Sri Lankan Tamil marriages. In J. Carsten, H. C. Chiu, S. Magee, E. Papadaki, & K. M. Reece (Eds.), *Marriage in past, present and future tense* (pp. 118–139). UCL Press.

Padilla, M. B., Hirsch, J. S., Munoz-Laboy, M., Sember, R. E., & Parker, R. G. (2007). Love and globalization. In *Transformations of intimacy in the contemporary.* Vanderbilt University Press.

CHAPTER 5

Migration, Reproduction, and the Demography of Citizenship

Abstract The notion of the *transnational making of population*, empirically identified in light of the Moroccan families' case study through the previous chapters, will be discussed in this chapter with reference to the Italian background. I argue that this context provides an emblematic window onto contemporary as well as contentious issues of nationality, population, and belonging. Italian society is undergoing significant demographic changes: on the one hand, the widespread decrease in birth rates and extended life expectancies lead to a substantial aging of the population; on the other hand, rising immigration rates constitute an important factor behind the renewal and growth of young population segments—a process that is also driven by the fertility and natality contribution of certain minorities, such as Moroccan communities. However, this population transformation is not embraced by the national boundaries of Italian belonging, as citizenship policies remain strictly fixed on the right of "blood" ("ius sanguinis"). In view of these opposing dynamics, this chapter concludes by framing in terms of *demography of citizenship* the growing gap between the progressive increase of non-citizen minors and children born to immigrants in Italy, on the one hand, and the available pathways for them to acquire the citizenship in the country where they were raised and educated, on the other hand.

Keywords Immigration in Italy • Italian demography • Demographic discourse • Second generation and citizenship • "ius sanguinis"

© The Author(s), under exclusive license to Springer Nature Switzerland AG 2024
F. Decimo, *Lives in Motion*, Palgrave Studies in Mediating Kinship, Representation, and Difference,
https://doi.org/10.1007/978-3-031-65583-8_5

107

The process of *transnational making of population* that Moroccan families have developed over time by moving and weaving stories of marriages and generations does not unfold in a socio-political vacuum. Rather, it takes place in Italy by entangling migrants' lives in a discursive, political, and juridical web that significantly connotes and forges the social dynamics at play. Family formation, demography, and the rule of law are indeed inter-related, and they give rise to a contentious political field ignited by issues of nationality, identity, and belonging. The following pages will delve into this field, shifting the focus from Moroccan migration per se to shed light on the broader social, political, and juridical environment in which it takes place. Specifically, on the one side, this chapter considers the quantitative features of the Moroccan immigration flow in comparison with the other main flows; on the other side, the analysis presented here takes into account the way the politics of nationality is pursued in Italy. Since I focus on the divergence between population movements and belonging in Italy, the following analysis addresses the Moroccan minority in comparison with other non-EU migrants in Italy, understood as minorities whose legal status is more vulnerable than that of EU ones.

My analysis proceeds by first outlining the aggregate effect of a multi-tude of family trajectories and reproductive choices, highlighting the fact that these population evolutions take place when the gender composition of the migratory flow undergoes a shift. Second, I consider how the inter-relation of migration and family formation intersects with Italian popula-tion transformations. I delve into this point not only in light of statistical data but also by pointing out the ethnonational meanings and implicit assumptions that permeate experts' population discourses. Finally, I focus on citizenship and its politics in Italy, drawing the attention to the way it remains strictly fixed on the right of "blood"—"ius sanguinis"—notwith-standing the significant transformation Italian society is undergoing with increasing numbers of immigrants and their descendants.

Moroccan Families Among Others: The Aggregate Effect of Intimate Choices

The multitude of migratory trajectories this book explores is part of a wider and crucially significant socio-demographic metamorphosis that has been sweeping over Italy since the late 1980s, that is, when the country shifted from being one of the most relevant countries of emigration in the world (1860–1985) to being an equally significant immigration destina-tion. Moroccan movements have figured prominently in Italy's history of immigration: they have not only been one of the main non-EU inflows since the first immigrants arrived (Fig. 5.1) but also generated an inner socio-demographic evolution that has triggered further changes. Specifically, I refer to the increasing participation of women in a movement that was initially overwhelmingly male: this transformation is manifested in the evolution of the Moroccan sex ratio in Italy, which shifted from four men to one woman in 1995 to almost one-to-one in current times (Fig. 5.2).

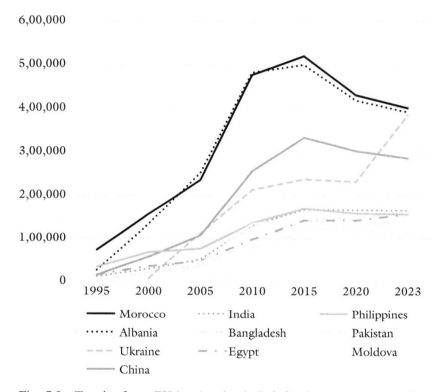

Fig. 5.1 Trends of non-EU immigration in Italy by the ten main nationalities (1995–2023). (Source: Ministry of Interior, residence permits (Data available on the following websites (last access, July 7, 2024): https://demo.istat.it/tavole/?t=permessi&l=it for the years 1995–2008; https://demo.istat.it/tavole/?l=it&t=noncomunitari for the years 2008–2021; and on Istat (2023a) for the 2023))

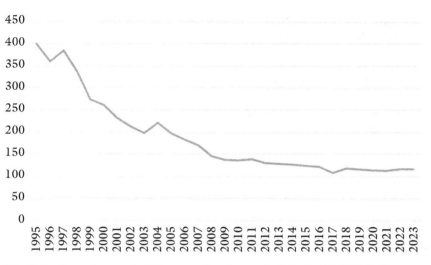

Fig. 5.2 Sex-ratio trend of Moroccan immigrants in Italy (1995–2023). (Source: Ministry of Interior, residence permits [Data available on the following websites (last access, July 7, 2024): https://demo.istat.it/tavole/?t=permessi&l=it for the years 1995–2008; https://demo.istat.it/tavole/?l=it&t=noncomunitari for the years 2008–2021; and on Istat (2023a) for the 2023])

This shift points to the scope of the myriad of female choices, household adjustments, and gender interplays explored in the previous chapters by which women have taken on more and more weight in the migratory process, in both qualitative and quantitative terms. As I investigate in more depth in Chap. 2, indeed, female participation in migration has not simply added numerically to male inflows; it has instead involved an evolution in settlement patterns from a sum of single individuals to an articulation of families and generations. Italian policies concerning residence permits and family reunification for non-EU migrants have favored this evolution: these policies are less strict than those in other EU countries, where criteria of cultural integration and bureaucratic procedures have become more stringent in terms of admitting family members (Ambrosini et al., 2014; Odasso, 2021). It is a fact that in 2022, 66% of residence permits for Moroccan immigrants were granted for familial reasons (ISTAT, 2023a, p. 2).

The Moroccan migratory flow is not the only one to have experienced such an internal transformation of its gender composition: the Albanian immigration flow shows a very similar trend, with 61% of residence

5 MIGRATION, REPRODUCTION, AND THE DEMOGRAPHY OF CITIZENSHIP   111

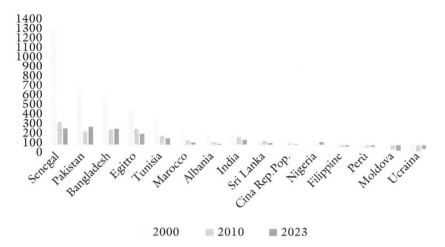

Fig. 5.3 Sex-ratio trend of immigrants in Italy by the 15 main nationalities (2000; 2010; 2023). (Source: Ministry of Interior, residence permits (Data available on the following websites (last access, July 7, 2024): https://demo.istat.it/tavole/?t=permessi&l=it for 2000; https://demo.istat.it/tavole/?l=it&t=noncom unitari for 2010; and on Istat (2023a) for 2023))

permits approved for family reasons in 2022 (ISTAT, 2023a); other flows (from Senegal, Pakistan, Egypt, Bangladesh, and Tunisia) remain predominantly male and entail lower rates of family reunification, although the numbers of family reunifications (25–40% of residence permits) and the decrease in the gender imbalance since 2000 have been significant (Fig. 5.3).

Most importantly, this gender turn brings about an equally significant generational upsurge, as Moroccan minority communities—along with the other, similar ones listed just above—are the non-EU minority to have brought more children into the world in Italy (Table 5.1).

It is worth framing and understanding these gender and generational evolutions through the lens of the migration-reproduction nexus. The relevance of this perspective becomes clear if we shift the focus of the analysis to the migratory flows that are composed predominantly of women, like the Filipino, Peruvian, Ecuadorian, Moldavian, and Uranian ones (Fig. 5.3). In these cases, migrant women's movement is driven by a demand for care work that represents a structural branch of the Italian labor market. These women are widely hired as around-the-clock domestic employees tasked with caring for disabled people or those who are not

Table 5.1 Numbers of children born in Italy of one or both foreign parents by the main 15 non-European nationalities (2022)

	Children born in Italy of at least a foreign parent. (Numbers)	Children of both foreign parents. (Percentages)
Morocco	9.965	68
Albania	9.168	63
Bangladesh	3.872	90
Nigeria	3.760	92
Egypt	3.045	80
India	2.825	83
Pakistan	2.715	79
China	2.124	84
Tunisia	2.060	63
Senegal	1.847	70
Moldova	1.752	65
Sri Lanka	1.234	91
Philippines	1.461	73
Ukraine	2.168	49
Peru	1.441	58

Source: Istat (2023b)

self-sufficient (Ambrosini, 2013a; Marchetti, 2022). The point I would like to stress here is that this female immigration does not entail household settlements or the formation of generations in Italy to any a significant extent (Ambrosini, 2015). In these circumstances, women instead forfeit their ability to have their own family lives here in Italy so as to maximize their capacity to produce monetary remittances and grant care and reproduction to their loved ones who remain at a distance. As the ratio between minors and adult women shows (Fig. 5.4), the chances of a woman living with her baby or child are disproportionally lower for migratory flows that remain highly selective by gender. This is true regardless of time since immigration, as can be seen from the low percentage of minors living with Filipino women, one of the migrant groups that has been in Italy the longest. On the other side, women whose migration is linked to male migration, that is, those who are part of migratory flows tending toward a more balanced gender composition, have higher odds of giving rise to second generations in Italy.

In conclusion, the *transnational population-making* process depicted thus far in light of Moroccan settlements in Italy pertains to other,

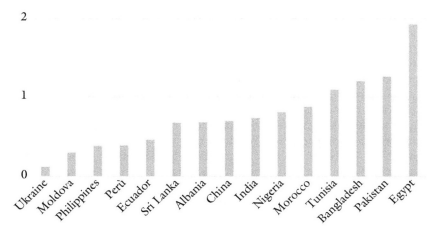

Fig. 5.4 Immigrant adult women-minors' ratio by the 15 main female nationalities in Italy (2021). (Source: Ministry of Interior, residence permits [Data available on the following website (last access, July 7, 2024): https://demo.istat.it/tavole/?l=it&t=noncomunitari])

selective migratory flows as well. The more mobility, marriage, and fertility are interrelated, the greater the demographic impact that immigration is destined to have in the long run. This is not a generalized dynamic: as revealed by the life courses and family stories charted in my analysis, for such a population evolution to take off, a specific combination of family stories, love affairs, and fertility preferences—namely, a specific intertwining of social capital and intimate patterns—must take place. As I argue in the following sections, however, whatever the cumulative demographic consequences generated by this sum of private choices, they are thoroughly overwhelmed by the discursive and political connotations attached to them.

Enacting High Fertility in a Low-Natality Country: A Controversial Lineage

In Italy, the gender and generational dynamics driven by migrant families outlined so far have prompted what Alba and Foner (2015, p. 14) define as a "diversity transition", that is, a shift in the national population involving an increase in the proportion of children born to immigrant parents as compared with other births. Data in this regard are significant: in 2022, one child in five has at least one foreign parent, while in 13% of births, both parents are foreign (Fig. 5.5). In the northern Italian regions, these rates are even higher, with 25–30% and 20% of children born to at least one and both foreign parents, respectively (Istat, 2023b).

This transition unfolds in a national context that is undergoing a significant demographic crisis: on the one hand, extended life expectancy is leading to substantial aging of the population; on the other hand, there has long been a widespread natality decrease, with deaths increasingly outnumbering births since 2003 (Fig. 5.6). The decreasing Italian fertility rate is the datum that underlies this gap: with the number of children per woman amounting to 1.18 in 2022, Italy has one of the lowest-low fertility rates countries in Europe (Kohler et al., 2002), well below the threshold that would guarantee a balanced generational turnover. The contribution of foreign woman[1] is particularly significant in this context,

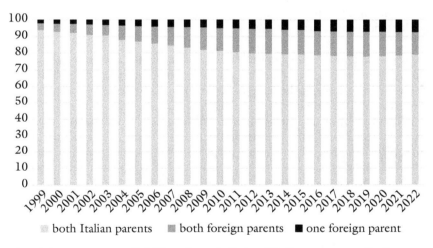

Fig. 5.5 Percentage of children born in Italy of Italian parents and of one or both foreign parents (1999–2022). (Source: Istat (2023b))

---

[1] Data on the fertility rates of foreign women in Italy are not available by nationality except for 2012, and then only for the main minority communities: see Giannantoni et al. (2018) in this regard.

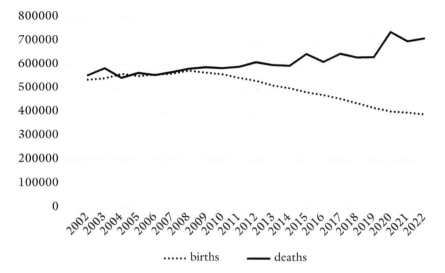

Fig. 5.6 Birth and death trends in Italy, 2002–2022. (Source: Births and deaths, Istat (Data available on the following websites (last access, July 7, 2024): https://demo.istat.it/app/?i=FE3&l=it for births; https://demo.istat.it/app/?i=ISM&l=it for deaths))

as their higher fertility rate (1.87) is sufficient to raise total fertility in Italy to a rate of 1.24 (Fig. 5.7).

Various components of Italian society—nationals and foreign, old and young people—are thus engaged in a demographic process of the social and political magnitude of which is becoming increasingly controversial. Similarly to other national contexts, indeed, demographic issues in Italy are also understood through frames that, far from being neutral and descriptive, are highly qualitative and judgmental in nature. These frames distort the definition of population far beyond that of a set of "individuals present at a given moment in a given place" (Le Bras, 2000, p. 12). Specifically, family and immigration issues—along with questions of gender—represent, here as elsewhere, the target of rising political attention fueled by right-wing and far-right parties and movements drawing more or less explicitly on "great replacement" conspiracy theories. This ideological frame has deep international and national historical roots. As Lucassen (2022) reconstructs, the "great replacement" conspiracy theory has been forged on a global scale through the story of Europeans' mass

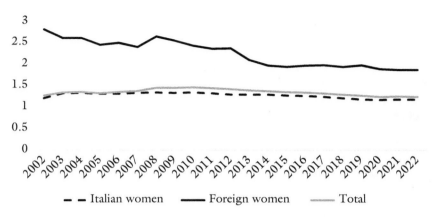

Fig. 5.7 Fertility rates in Italy, Italian and foreign women, 2004–2022. (Source: Fertility rates, Istat (Data available on the following website (last access, July 7, 2024): https://demo.istat.it/app/?i=FE1&l=it))

immigration to the United States starting in the 1830s. The discursive elements of this ideology—nativism, racism, siege, and conspiracy—have been constructed and affirmed through a rhetoric that systematically represents the supremacy of imperiled national ingroups against the degenerative inferiority of dangerous outgroups. Such a narrative structure was reignited and updated after 9/11 in light of the more recent "clash of civilizations" current of thought, finding new fuel on the other side of the Atlantic. It has spread in Europe, advancing the idea that a "Eurabian" order is threatening to impose itself on liberal societies. In this context and the Italian political arena as well, the idea that immigration represents a threat to the very survival of the national population and identity has crept further and further into representations and debates. According to this frame, Italian right-wing parties in particular have changed the language and contents of their discourse in the last decade, focusing more and more frequently on issues of racial identity and purity to advance their political and electoral campaigns (Pellegrino, 2021; Petrovich Njegosh, 2023; Quassoli, 2025).

In reality, on closer inspection, the roots of the intersection of demographic and political issues in Italy extend back to the Fascist era when the state established a distinct emphasis on the national population as a racial concern. This era witnessed not only pro-natalist initiatives central to regime policies, as outlined by Ipsen (1996), but also the delineation of the family as a social and political institution in the 1942 Civil Code. Such policies were underpinned by a conception of the national population viewed through the lens of eugenics. This amalgamation of demography and medicine, statistics and hygiene, public policy, and propaganda spawned the notion that reproduction could be managed, selectively controlled, and purged of undesirable traits. Various strategies were employed, from the exaltation of reproduction and the maternal role as described by Saraceno (1991) to the establishment of specific initiatives and institutions, notably the Manifesto of the Racial Scientists and the Department of Demography and Race, as highlighted by Cassata (2006).

Following the fall of the Fascist regime and advent of the Italian Republic, the structures and institutions of these demographic and racial policies were gradually dismantled. Nevertheless, the issue of population—encompassing both its size and its composition—retained political significance over time, prompting governments of varying political persuasions to advocate for demographic policies. Consequently, albeit within changing historical and institutional contexts, Italians' family affairs and reproductive sphere have continued to constitute focal points of public interest. Notably, contemporary politics, under the guise of addressing the demographic crisis, has garnered renewed legitimacy by delving into matters of intimate preferences, gender, and household formation, with a particular emphasis on women's reproductive choices and bodily autonomy that also extends to debates surrounding abortion rights (Krause, 2001; Marchesi, 2016).

This heightened focus targets predominantly Italian couples and women, imagined respectively as heterosexual and of native origins, encouraging them to reproduce. Conversely, the reproductive decisions of immigrants and non-heteronormative couples are subjected to increasing scrutiny, interrogation, and opposition at the levels of both media discourse and public policy. Within this milieu, demographic scientists are continually called upon to contribute analyses and forecasts. However, the history of Italian scientific institutions' entanglement with past national demographic politics, particularly during the Fascist era, complicates matters quite a bit. Despite contemporary demographic studies being disassociated from this past, they often perceive population issues through a political lens and tend to employ evaluative and biased categories. Notably, immigration is frequently construed as a national question in which migrants are viewed as either a potential, selected resource or a threat, reflecting a recurring interpretative framework (Etzemüller, 2011).

While acknowledging the role of immigrants in compensating for demographic imbalances, particularly within the labor market, analysts, and population experts remain hesitant or biased in recognizing immigrants as a structural component of the Italian population (Ambrosini, 2013b; Colombo & Dalla-Zuanna, 2019). Even considerations of immigration and foreign fertility as potential solutions to demographic challenges are overshadowed by persistent narratives framing the natality issue as the exclusive responsibility of Italians. This discourse subtly delineates an "us" versus "them" dichotomy, emphasizing the need for Italians to reproduce so as to prevent an influx of foreigners (Decimo, 2015). This dichotomy becomes more pronounced when forecasts predict an increase in immigrant numbers beyond what are perceived as sustainable levels. The prevailing discourse in Italy adopts a language of threat and selection, rejecting the notion that population imbalances could be rectified by the reproductive contributions of foreigners.[2] High-level political and scientific exponents (Valditara et al., 2016), through reasoning that definitively entangles demographic thought with political and demographic concerns, argue that Italian society needs to prevent the risk of degeneration posed by immigration.[3] Specifically, they point to immigration from Islamic countries as part of a broader hegemonic plan of ethnic substitution supposedly promoted by foreign Muslim governments and movements worldwide, not only by adopting religious and ideological levers that impede second-generation assimilation but also especially through their higher fertility rates. They warn, therefore, that an only apparently peaceful invasion is taking place in Italy, the same kind of invasion they claim already irretrievably overwhelmed the classic world as well as the recent past of northern European countries.

Despite themselves, Moroccan migrants are the main (although not the only) actors in these processes: they are the main non-EU minority in Italy, Muslim, oriented toward family settlement, and engendering higher fertility rates than the Italian average. Their intimate choices, as reconstructed through the qualitative, in-depth investigation presented in the previous chapters, are destined to be harnessed to this web of discourses

---

[2] For a more exhaustive review of the way demographic discourse frames immigration in Italy, see Decimo (2015).

[3] I refer here to the theses argued by Giuseppe Valditara, Gianandrea Gaiani, and Gian Carlo Blangiardo (2016), respectively, the Education and Merit Minister since 2022, consultant for the Defense and Interior Ministries on security matters since 2018, and president of ISTAT (National Institute of Statistics) from 2019 to 2023.

and policies. The affective bonds these migrants tie and untie across migration, who they marry, the households they form, the babies they bring into the world, and the daily lives they conduct with their loved ones: all these life events are fostered as a broad, transnational momentum into the future. However, when they land in the national context of Italy, when the intergenerational dynamics they engender seek a place within the boundaries of the Italian population, the anchors they turn to prove unstable and slippery. A gap begins to open up between the significant demographic process driven by their migration (along with that of many other men and women of different origins) and the way it is understood and narrated, both at the level of demographic analyses and in the arena of media and political discourse. And yet the story does not remotely end here: there is a further dimension that lies at the heart of this gap, a dimension stemming from the way Italian citizenship is defined and outlines the boundaries between ancestry, lineage, and national belonging, as I will show in the following analysis.

"Blood" Law and the Boundaries of Belonging in Italy

Although Italy has been an immigration country for several decades now, citizenship policies remain firmly and primarily tied to the principle of "ius sanguinis" or "right of blood". No matter where they are born, individuals are considered Italian if they have one Italian parent or an ancestor who never renounced his or her Italian citizenship (Zincone, 2006). A track for the naturalization of foreigners who reside stably in the country is available as well, but this is a complicated process without guaranteed success, especially compared to the highroad accessible to those who can claim Italian lineage. Such a preferential access to citizenship through ancestry is a product of the country's long history of emigration and the extensive Italian diaspora (Gabaccia, 2000). Embedded within the significance of this collective memory, however, there are also political motivations for embracing this "Italy outside of Italy" as an asset abroad of current economic and political interest to the country (Zincone, 2006). Through legislation[4] that supports not only instances of Italian citizenship by descendants but also the recovery of lineage by foreigners who claim to have at least an Italian ancestor (Zincone & Basili, 2010), from 1992

---

[4] Law n. 91, February 5, 1992.

120   F. DECIMO

onwards, a conspicuous number of subjects have been entitled to request Italian citizenship (Gallo & Tintori, 2006; Tintori, 2011; Zincone, 2006). Furthermore, this program to foster the acquisition of nationality by people identified as potentially Italian was associated with the establishment of the right to dual nationality, thanks to which applicants were enabled to preserve their original national rights, thereby avoiding facing any possible dilemma. Tintori (2009, 2011) has empirically reconstructed how the idea of transforming Italian descendants abroad into nationals came to be, focusing his research in Latin America and retracing the network of lobbies working from Italy to shape this new national/international arena. On the one side, he considers such politics in terms of new voters and their influence on the Italian electoral arena; on the other side, Tintori (2011) highlights how new rights of mobility have been granted in this way, allowing a million Latin Americans to secure Italian passports between 1998 and 2008.

However, the impact of the 1992 citizenship law with its stress on bloodline becomes even more pertinent if we consider how it affects the population dynamics currently playing out in Italy. By contrast, in fact, immigration stands out as an established phenomenon in Italy in this same period, clearly preluding the profound transformations of the country's socio-demographic composition (Zincone & Basili, 2010). For these immigrants and their descendants, citizenship can only be obtained by request, on the basis of residency or marriage. Naturalization through residence in particular is the only option available to most immigrants and is subjected to significant constraints: non-EU nationals aiming to become naturalized in Italy are required to prove ten years of continuous residence (while for refugees and EU citizens, the residence length requirement is five and four years, respectively). In addition, individual or household income must be documented as a guarantee of stability and self-sufficiency, along with the absence of a criminal record or any other element considered potentially incompatible with the public interest. Naturalization requests based on residence are supposed to be processed in two years, but in reality, the process usually takes twice as long and involves a high degree of discretion (MPG, 2013): the issue, indeed, is that naturalization continues to be conceptualized as a concession rather than a right individual can claim upon fulfilling certain requirements (Locchi, 2014, p. 484).

Most importantly, the naturalization track for minors and individuals born in Italy is similar to the adult one. Immigrant minors can be conferred Italian citizenship by transmission if a parent acquires it, while individuals born in Italy to foreign parents can request it by election when they have reached their legal majority (18 years old) but not yet turned 20. They also need to meet other strict conditions: like adult immigrants, applicants born in Italy must also have maintained documented status at all times; this means being born to parents who hold residency permits and being able to prove uninterrupted legal residence in Italy (Tintori, 2013). The processing of these citizenship requests takes an extended period of time, no different from the timeframe for adults, and they are similarly understood as grants instead of entitlements. In a nutshell, being born or raised in Italy does not represent a decisive factor for large numbers of second-generation youths in their path to acquiring Italian citizenship.[5]

Overall, the ordinary track to naturalization in Italy, considered in comparison with most European countries, is more restrictive and presents more serious obstacles in terms of proof of income, residence requirements, bureaucratic procedures, lack of assistance, length of waiting times, and the opaqueness and/or accountability of the decision-making process (MPG, 2013, pp. 11–12). This is particularly true for immigrants' children, even in comparison with Spain whose naturalization regime is similar (Finotelli et al., 2018). The Italian politics of citizenship thus encompass "blood" law and naturalization, emigration and immigration, all in the same framework. As such, they delineate an uneven, segmented landscape of belonging/exclusion in which lineage—that is, having Italian ancestry—plays a pivotal, privileged role. On the one side, the descendants of Italian emigrants—even several generations removed, who might not speak Italian or have ever stepped foot in the country—have concrete opportunities to obtain Italian citizenship, while, on the other side, the children of immigrants who were born and raised in Italy are considered foreigners.

The data in this regard are significant: Moroccans and Albanians stand out in terms of their high number of citizenship acquisitions, but this figure is due to the conspicuous and long-term settlement of these minorities in Italy. On the other hand, it is worth noting that the number of Brazilians and Argentines who acquired Italian citizenship is much higher than that of other foreign minorities historically present in Italy. In particular, it is important to stress that these individuals obtained Italian citizenship by

---

[5] For more references to the debate and political activities around the reform of the "ius soli" ("right of the soil" or birthright citizenship) provisions in the Italian national context, see Locchi (2014) and Tintori (2018).

lineage in 77% and 92% of cases, respectively (Table 5.2), basically by claiming an Italian ancestor. Indeed, given the statistics of these immigrants and their children in Italy, neither transmission through the naturalization of parents nor election could support such high numbers of citizenship acquisitions.

We can obtain an even clearer view of the weight of citizenship by "blood" in shaping—in a hierarchical way—the way immigration and Italian nationality fit together by focusing on the naturalization rate based on long-term residence permits (Table 5.3), that is, permits based on qualifications similar to those required to apply for citizenship.[6] In this

Table 5.2 Percentages of Italian citizenship acquisition through transmission from parents, lineage (ius sanguinis) or choice (18 y/o) and total numbers of Italian naturalization by the main 20 nationalities of origin (2021)

	Transmission, lineage, or choice (%)	Total numbers of Italian naturalization
Argentina	92	3.669
Brazil	77	5.460
Egypt	60	3.531
Pakistan	59	4.410
Nigeria	56	2.198
Bangladesh	55	5.116
Philippines	55	2.342
Sri Lanka	54	1.608
Tunisia	52	3.036
Morocco	48	16.588
Senegal	47	2.881
Kosovo	47	1.718
India	43	4.489
North Macedonia	42	2.718
Albania	39	22.493
Romania	36	9.435
Ecuador	34	3.362
Perù	32	2.748
Moldova	25	3.633
Ukraine	22	2.682
Total	46	121.457

Source: Unpublished data provided on request by the Istat Contact Centre (2023)

[6] To obtain long-term residence, applicants are required to prove five years of continuous residence in Italy (Decree 3/2007 transposing 2003/109/EC). In fact, this long-term residence permit represents a premise for naturalization (Decree 572/1993, State Council 799/1999).

Table 5.3 Italian naturalization rate[a] by the main ten nationalities of origin (2021)

Argentina	*63*
Brazil	18
Albania	8
Pakistan	7
Bangladesh	6
Morocco	6
India	4
Moldova	4
Egypt	4
Romania	1

[a]Rate calculated as the ratio of the total number of citizenships granted over the long-term residence permits (Finotelli et al., 2018)

Source: Ministry of Interior, long-term residence permits and citizenship concessions (Data available on the following websites (last access, July 7, 2024): http://dati.istat.it/Index.aspx?DataSetCode=DCIS_PERMSOGG1 for long-term residence permits and http://stra-dati.istat.it/ for citizenship concessions)

case, opportunities to become Italian citizens are disproportionally distributed, with higher rates for Brazilians (18%) and Argentinians (63%), while all the other foreign minorities in Italy have much lower naturalization rates (3–8%).

On the other side, citizenship request denials confirm this pattern: citizenship applications are rejected at a higher rate for immigrants from Africa, Asia and non-EU Europeans (Fig. 5.8). Specific data on applicants' nations of origin are not available, but it is a fact that these are the continents of origin for the main minorities to have historically settled down in Italy—specifically, as mentioned above, Moroccans and Egyptians (Africa), Pakistan, India, and Bangladesh (Asia), as well as Moldavia and Albania (non-EU Europeans).

In light of these numbers, it is clear that the gap between "blood" law and naturalization in the Italian citizenship regime involves not only the level of legal principles but also the formal outline of national boundaries (Zincone, 2010). Rather, citizenship politics operate like a frame the Italian state imposes to shape its ideal national composition, drawing boundaries without regard for the actual collectivities who settle down, produce, and reproduce on its soil, that is, regardless of Italy's actual society. In this landscape, through population movements and particularly the *transnational making of population* I have traced in relation to Moroccan migrations, children are brought into the world whose trajectories and belonging unfold in Italy but who, despite the numbers, remain framed as foreigners, as out of place, accidental, undesired occurrences.

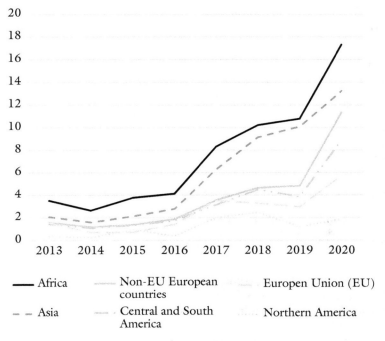

Fig. 5.8 Trends of citizenship rejection rate by applicants' continent of origin (2013–2020). (Source: Ministry of Interior [Data available on the following website (last access, July 7, 2024): http://www.libertaciviliimmigrazione.dlci.interno.gov.it/it/documentazione/statistica/cittadinanza])

Conclusion: Transnational Migrations, Generation, and the Demography of Citizenship

Opportunities for family settlement are quite unevenly distributed among migrants in Italy. Many migratory flows remain highly selective by gender, animated mainly by the search for work, and over time even replicate mobility chains based on labor. The chance to form a couple, set up enduring households and raise children in the context of immigration is indeed generated by an array of circumstances and processes that are not at all generalized. Rather, such circumstances and processes are embedded in practices and patterns which are unequally distributed across migratory circuits, bounded by ethnonational belonging and networks (Portes & Sensenbrenner, 1993; Wimmer, 2013).

Moroccan migration to Italy, quite similarly to Albanian inflows, has long been expanding to the point that these immigrants now constitute the main non-EU minority in Italy. The driving factors behind this trend are the progressive balancing of the sex ratio—that is, rising female participation in the migratory process—and the increasing settlement of families. Most importantly, this dynamic has engendered conditions in which Moroccan families have the highest natality rates among non-EU immigrants in Italy. In light of these data, the impact of the Moroccan transnational making of population, as qualitatively explored across the previous chapters, has become quantitatively significant in the Italian context. By drawing affective resources from Morocco, families and generations take root on the other side of the Mediterranean. This provides clear evidence of the strength of transnationalism: intimate choices, household articulation, and migration give rise to interwoven processes vital enough to generate a demographic dynamic unfolding between Morocco and Italy. Individual life courses and family cycles evolve by adopting this horizon of mobility, a horizon that clearly does not end at national borders but rather extends beyond them.

On deeper and more contentious levels, however, nation does matter. As scholars have thoroughly underscored, even in moments of peak transnationalism and global migration, these flows are enacted hand in hand with the enforcement and implementation of state institutions' power to delimit and categorize national populations, rather than diminishing such powers (Bauböck, 1998, 2003; Joppke, 1998; Zolberg, 1999). Migrants, for their part, do not remotely experience trajectories of boundless deterritorialization—they instead pursue the chance to act, as much as possible, as competent citizens in multiple national contexts (Faist, 2000; Kivisto, 2001; Waldinger, 2015; Waldinger & Fitzgerald, 2004).

Today's trends of border fortification and restrictive immigration policies have exacerbated these processes (Carens, 2013) and are mirrored by the construction of identity-based boundaries which take place more and more frequently within nations as well, threading through the national population itself and reifying the diversity of its composition (Decimo, 2015; Decimo & Gribaldo, 2017). The result of these dynamics is the forging of internal, domestic outsiders (De Genova, 2005; Dreby, 2015; Ngai, 2014) whose status ends up being jeopardized by public discourses, processes of racialization, and moves to undermine their standing under the law.

The Italian case, scrutinized in light of Moroccan immigration, eloquently illustrates these processes and more. On the one side, the demographic contribution of these families in terms of new births, in a country affected by dramatic demographic crisis for several decades now, is overlooked or even written off entirely. Viewed through the recent conspiracy-based lens of replacement theory that, in Italy as elsewhere, has been nourished by a new alliance between right-wing politicians and scientists, the presence and reproduction of people hailing from predominately Muslim countries represents a threat to Italian identity distinctly represented as Christian and white. On the other side, a stricter barrier is operating to outline the Italian population in keeping with these racial contours. The barrier in this case is constituted by citizenship, the status that represents the utmost mark of belonging to the body of the nation and, in the opposite direction, the most effective instrument for excluding people from it (Brubaker, 2009; Mann, 2005; Wimmer, 2002). The Italian system is strongly based on citizenship by "blood": lineage represents the high road for obtaining nationality, granted not only to the children of Italian parents—indeed, taking on one's parents' citizenship is a universal right (Bauböck, 2018)—but also to foreigners, born and raised abroad, who are able to claim an Italian ancestor, however distant. Naturalization by residence involves a much more tightly controlled path. Most importantly, being born and socialized in Italy is not a basis for acquiring Italian citizenship:—the disconnect is so great, in fact, that these individuals come of age and enter adulthood as foreigners in the country that, for all intents and purposes, they experience as their homeland. The way immigration policies and citizenship underpin the selective and stratified construction of nationality has been extensively investigated by scholars (Joppke, 2005; Zolberg, 2006), and Italy represents an eloquent case in this regard.

However, the point I wish to stress through my analysis is that these politics do not take place merely at the level of the legal principles and normative criteria defining nationality. Far from it: such politics instead have tangible downstream effects, affecting social processes that are already broadly and fully enacted through the deployment of a sum of non-reversible human lives (Dreby, 2015; Ngai, 2014). Considering the number and origin of individuals who are granted Italian nationality, naturalization rates, and the rejection rates for citizenship applications, the key point here is that the Italian state's chosen policy favors foreigners with imagined national origins (Anderson, 1991) but who may have no actual connection with Italy over generations of individuals who are born, educated, and rooted in the country. In so doing, the definition of who has right to national membership and who does not constitute more than a formal status; rather, it is an effective mold that the state uses to shape and elect a population, cutting out and discarding the components seen as lying outside its identity-based boundaries even though these people are born, raised, and living in Italy as well as participating fully in Italian society. In this perspective, the result is a *demography of citizenship*, by which I mean the social distribution of nationality in the resident population, considering not only the existing population but also the one in the making. This focus is reminiscent of the concept of civic stratification (Morris, 2002) but differs in that, here, I attempt to incorporate a specific consideration of how the population develops over time, the way ethnonational boundaries affect the way people are counted (Kertzer & Arel, 2002; Simon et al., 2015), and the political weight assigned to such boundaries (Alba, 2020). By curbing or preventing the recognition of migrants and their descendants as full members of the national project, certain ideas of identity, generation, descendance, diversity, population, persistence, and disappearance are nurtured while others are neglected or suppressed. This is a demographic politics that widens the gap between the imagined nation and the actual society. At the same time, such a process inevitably confers new political significance on the reproductive sphere: from being defined by the intimate choices and bonds involved in childbirth and -rearing, reproduction becomes a means for perpetuating the regimentation of the national process and, as such, a matter of increasing political interest and state governance.

REFERENCES

Alba, R. (2020). *The great demographic illusion: Majority, minority, and the expanding American mainstream*. Princeton University Press.

Alba, R., & Foner, N. (2015). *Strangers no more: Immigration and the challenges of integration in North America and Western Europe*. Princeton University Press.

Ambrosini, M. (2013a). *Irregular migration and invisible welfare*. Palgrave.

Ambrosini, M. (2013b). Immigration in Italy: Between economic acceptance and political rejection. *Journal of international migration and integration, 14*, 175–194.

Ambrosini, M. (2015). Parenting from a distance and processes of family reunification: A research on the Italian case. *Ethnicities, 15*(3), 440–459.

Ambrosini, M., Bonizzoni, P., & Triandafyllidou, A. (2014). Family migration in Southern Europe: Integration challenges and transnational dynamics: An introduction. *International Review of Sociology, 24*(3), 367–373.

128   F. DECIMO

Anderson, B. (1991). *Imagined communities: Reflections on the origin and spread of nationalism*. Verso.

Bauböck, R. (2018). Ius filiationis: A defence of citizenship by descent. *Debating Transformations of National Citizenship*, 83–89.

Bauböck, R. (1998). The crossing and the blurring of boundaries in international migration: Challanges for social and political theory. In R. Bauböck & J. Rundell (Eds.), *Blurred boundaries: Migration, ethnicity, citizenship* (pp. 17–52). Ashgate.

Bauböck, R. (2003). Towards a political theory of migrant transnationalism. *The International Migration Review, 37*(3), 700–723.

Brubaker, R. (2009). *Citizenship and nationhood in France and Germany*. Harvard University Press.

Carens, J. (2013). *The ethics of immigration*. Oxford University Press.

Cassata, F. (2006). *Molti, sani e forti. L'eugenetica in Italia*. Bollati Boringhieri.

Colombo, A. D., & Dalla-Zuanna, G. (2019). Immigration Italian style, 1977–2018. *Population and Development Review*, 585–615.

De Genova, N. (2005). *Working the boundaries: Race, space, and "Illegality" in Mexican Chicago*. Duke University Press.

Decimo, F. (2015). Nation and reproduction: Immigrants and their children in population discourse in Italy. *Nations and Nationalism, 21*(1), 139–161.

Decimo, F., & Gribaldo, A. (Eds.). (2017). *Boundaries within: Nation, kinship and identity among migrants and minorities* (Vol. 24). Springer.

Dreby, J. (2015). *Everyday illegal: When policies undermine immigrant families*. University of California Press.

Etzemüller, T. (2011). The population discourse: A transnational matrix. *Historical Social Research, 36*(2), 101–119.

Faist, T. (2000). Transnationalization in international migration: Implications for the study of citizenship and culture. *Ethnic and Racial Studies, 23*(2), 189–222.

Finotelli, C., La Barbera, M., & Echeverría, G. (2018). Beyond instrumental citizenship: The Spanish and Italian citizenship regimes in times of crisis. *Journal of Ethnic and Migration Studies, 44*(14), 2320–2339.

Gabaccia, D. (2000). *Italy's many diasporas*. Routledge.

Gallo, G., & Tintori, G. (2006). Come si diventa cittadini italiani. In G. Zincone (Ed.), *Familismo Legale. Come (non) Diventare Italiani*. Laterza.

Giannantoni, P., Ortensi, L. E., Strozza, S., & Gabrielli, G. (2018). La Fecondità degli Stranieri. In *Vita e Percorsi di Integrazione degli Immigrati in Italia* (pp. 101–121). Istat.

Ipsen, C. (1996). *Dictating demography: The problem of population in fascist Italy*. Cambridge University Press.

ISTAT. (2023a). *Statistiche Report. Cittadini non comunitari in Italia. Anni 2002–2023*, Roma. Retrieved March 15, 2024, from https://www.istat.it/it/archivio/289255

ISTAT. (2023b). *Statistiche Report. Natalità e fecondità della popolazione residente. Anno 2022, Roma.* Retrieved July 7, 2024, from, https://www.istat.it/comunicato-stampa/natalita-e-fecondita-della-popolazione-residente-anno-2022/

Joppke, C. (1998). Immigration challenges the nation-state. In C. Joppke (Ed.), *Challange to the nation-state. Immigration in Western Europe and the United States* (pp. 5–46). Oxford University Press.

Joppke, C. (2005). *Selecting by origin: Ethnic migration in the liberal state.* Harvard University Press.

Kertzer, D. I., & Arel, D. (Eds.). (2002). *Census and identity: The politics of race, ethnicity, and language in national censuses* (Vol. No. 1). Cambridge University Press.

Kivisto, P. (2001). Theorizing transnational immigration: A critical review of current efforts. *Ethnic and Racial Studies, 24*(4), 549–577.

Kohler, H. P., Billari, F., & Ortega, J. A. (2002). The emergence of lowest-low fertility in Europe during the 1990s. *Population Development Review, 28*(4), 641–680.

Krause, E. L. (2001). "Empty cradles" and the quiet revolution: Demographic discourse and cultural struggles of gender, race, and class in Italy. *Cultural Anthropology, 16*(4), 576–611.

Le Bras, H. (2000). *L'Invention des populations. Biologie, idéologie et politique.* Éditions Odile Jacob.

Locchi, M. C. (2014). Lo ius soli nel dibattito pubblico italiano. *Quaderni costituzionali, 34*(2), 483–506.

Lucassen, L. (2022). *Roots of a murderous idea: "Replacement" thinking in the Atlantic World since the early 19th century* (Research Paper No. 55). International Institute of Social History.

Mann, M. (2005). *The dark side of democracy: Explaining ethnic cleansing.* Cambridge University Press.

Marchesi, M. (2016). Reproducing Italians: Contested biopolitics in the age of 'replacement anxiety'. In *Reproduction and Biopolitics* (pp. 35–52). Routledge.

Marchetti, S. (2022). *Migration and domestic work.* Springer.

Morris, L. (2002). *Managing migration: Civic stratification and migrants' rights.* Routledge.

MPG (Migration Policy Group). (2013). *Access to citizenship and its impact on immigrant integration. Handbook for Italy.* EUI.

Ngai, M. M. (2014). *Impossible subjects: Illegal aliens and the making of modern America.* Princeton University Press.

Pellegrino, D. (2021). Teorie cospirazioniste demografiche. In Pellegrino, D. & Pannofino, N. (a cura di), *Trame nascoste: teorie della cospirazione e miti sul lato in ombra della società* (pp. 9–27). Mimesis.

Petrovich Njegosh, T. (2023). La teoria della sostituzione etnica in Italia: una narrazione razzista e sessista. *From the European South, 12,* 105–122.

Portes, A., & Sensenbrenner, J. (1993). Embeddedness and immigration: Notes on the social determinants of economic action. *American Journal of Sociology, 98*(6), 1320–1350.

Odasso, L. (2021). Negotiating legitimacy: Binational couples in the face of immigration bureaucracy in Belgium and Italy. *Anthropologica, 63*(1), 1–30.

Quassoli, F. (2025). 'At Risk of Extinction'. Immigration, national identity, and global class conflict according to 'Population Replacement Conspiracy Theory' in Contemporary Italy. In G. Navarini (Ed.), *Conspiracy theories in contemporary Italy. Origins and uses of a cultural and political production.* Routledge.

Saraceno, C. (1991). Redifining maternity and paternity: Gender, pronatalism and social policies in fascist Italy. In G. Bock & P. Thane (Eds.), *Maternity and gender policies. Women and the European welfare state 1880s–1950s.* Routledge.

Simon, P., Piché, V., & Gagnon, A. (Eds.). (2015). *Social statistics and ethnic diversity. Cross-national perspectives in classification and identity politics.* Springer.

Tintori, G. (2009). *Fardelli d'Italia?: conseguenze nazionali e transnazionali delle politiche di cittadinanza italiane.* Carocci.

Tintori, G. (2011). The transnational political practices of "Latin American Italians". *International Migration, 49*(3), 168–188.

Tintori, G. (2013). *Naturalisations procedures for immigrants EUI.* Robert Schuhman Centre of Advanced Studies.

Tintori, G. (2018). Ius soli the Italian way. The long and winding road to reform the citizenship law. *Contemporary Italian Politics, 10*(4), 434–450.

Valditara, G., Blangiardo, G. C., & Gaiani, G. (2016). *Immigrazione. Tutto quello che dovremmo sapere.* Aracne.

Waldinger, R. (2015). *The cross-border connection. Immigrants, emigrants, and their homelands.* Harvard University Press.

Waldinger, R., & Fitzgerald, D. (2004). Transnationalism in question. *American Journal of Sociology, 109*(5), 1177–1195.

Wimmer, A. (2002). *Nationalistic exclusion and ethnic conflict.* Cambridge University Press.

Wimmer, A. (2013). *Ethnic boundary making.* Oxford University Press.

Zincone, G. (2006). *Familismo legale. Come (non) diventare italiani.* Laterza.

Zincone, G. (2010). *Citizenship policy making in Mediterranean EU states: Italy.* Robert Schuhman Centre of Advanced Studies.

Zincone, G., & Basili, M. (2010). EUDO citizenship observatory. *Country report: Italy. EUI, Florence: Robert Schuhman Centre of advanced studies.*

Zolberg, A. R. (1999). Matters of state: Theorizing immigration policy. In C. M. Hirschman, P. Kasinitz, & J. DeWind (Eds.), *The handbook of international migration: The American experience* (pp. 71–93). Russel Sage Foundation.

Zolberg, A. R. (2006). *A nation by design: Immigration policy in the fashioning of America.* Harvard University.

CHAPTER 6

Conclusion: Lives in Motion and Their Future

Abstract This conclusive chapter of *Lives in motion* aims to move beyond the conclusions, shoring up each individual chapter of the book in order to pose a final reflection hinged on three consequential arguments, concerning: (1) The role transnational marriages play in sustaining a process of population-making that crosses national borders; (2) The circulation of affective resources as social capital that advantages migrants to a different extent in family settlement efforts; and (3) The structure of reproductive opportunities that selectively underpins migrants' movement and the need to rethink how demography and nationality are entangled.

Keywords Kinship and migration • Transnational marriage as social capital • Reproductive opportunities • Demography and nationality • Population and social change

The arguments developed throughout this book took inspiration from the need to reconsider how family life takes place in controversial times of global migration, increasing separation, and restrictive mobility policies. The research approach I adopted was aimed at problematizing the circumstances that enable migrants to achieve their own family fulfillment, understanding this as a process only partially represented by opportunities for household reunification. More in-depth, the study I have carried out seeks

© The Author(s), under exclusive license to Springer Nature Switzerland AG 2024
F. Decimo, *Lives in Motion*, Palgrave Studies in Mediating Kinship, Representation, and Difference,
https://doi.org/10.1007/978-3-031-65583-8_6

131

to explore the underlying dynamics of family formation across migration: this perspective has entailed analyzing how social reproduction is restructured according to the times and spaces of transnationalism, and the respective practices, conceptions, and occurrences involved. I was interested in understanding how individuals, households, and kinship groups conceive and accomplish family perpetuation as a process that draws on social resources in the context of origin but takes place abroad, away from where the home once was. The analysis has been based on life courses collected through the fieldwork I conducted among Moroccan families in Italy. Specifically, I charted how mobility and family formation intertwined from the decision to move to the chances of finding a partner, the choice to settle down together in the context of immigration, and the possible subsequent births that these moves give rise to.

As I conducted my research, I was conscious that the unfolding of family events and the process underpinning them cannot ever be taken for granted across migration: the interrelation of movement and family-making do indeed constitute a highly politicized issue in times of border enforcement, contested mobility, and uncertain settlement. This is true both because family movements are considered by states as forms of entry that might elude today's restrictive immigration policies, and because the establishment of migrants' households entails demographic issues that call into play identity and the national boundaries of populations. This latter point is the one I have delved into more deeply throughout my analysis, bearing in mind that migration, demography, and nationhood drive the winds of political flags that are blowing to an increasing extent around the world, exacerbating identity-based categories and bordering processes. The Italian context, firmly attached to conferring citizenship on the basis of blood, represents an eloquent case study in this regard.

These lines of reasoning have been laid out, setting off from an analysis of the marriage and migration nexus (Chap. 2); then reconstructing how family formation and fertility take shape in the context of immigration (Chap. 3); considering the way intimacy and the private sphere are experienced (Chap. 4); and finally, by reconstructing in a bird's eye view the aggregate impact that a sum of personal choices can have from the perspective of national demography and citizenship politics (Chap. 5). Besides the conclusions shoring up the different chapters of the book, my purpose here is to disentangle and stress three main points stemming from the different aspects discussed throughout my analysis so as to offer an overall, final reflection on *Lives in motion*:

Marriages, Kinship, and the Transnational Making of Population

Transnational marriages feature a pivotal knot that underpins the evolution of Moroccan mobility networking in Italy, as it has developed over time after an initial migratory chain of predominately single men. Indeed, the life courses examined in my research clearly show that a marriage arena has been deployed between the two sides of the Mediterranean, thereby enabling migrants to find the right mate with whom to start a family in the context of immigration. As I have delved into with Chap. 2, this arena has been shaped by the multiplying social relationships that migrants forge and cultivate through their back and forward trips and the increasing involvement of kin in their migratory trajectory. Relatives do indeed actively participate in individuals' search for a spouse, possibly also sponsoring candidates when the circumstances are right to enter into a convenient marriage agreement. Female participation in this system has increased: women appreciate the idea of marrying a migrant spouse and establishing themselves in Italy, considering this an opportunity for mobility and a legitimate way to avoid patrilocal norms. Overall, the cumulative effect of individual agency and kinship networking has engendered a transnational system that strengthens the understanding of marriage as social capital and not only an individual choice. On one side, the accounts I collected make evident that the opposition between arranged marriages and marriages by choice represents a false dichotomy: in any event, relatives are involved in the process of conjugal matching, and they also support individuals in case of conjugal failure, as highlighted in Chap. 4. On the other side, such a wide field of relationships grants ease, confidence, and speed in promoting marriages and facilitating family formation: individuals who make use of kinship and personal networks have a good chance of turning their lives from single to married simply and quickly.

In so doing, kinship not only gives rise to networks that favor individuals in their mobility project, as stressed by recent studies (Andrikopoulos, 2023). Through a broader mobilization of affective and social resources, kinship literally drives the relocation of consuetudinary spheres of social reproduction abroad. Viewed from this perspective, the salience of kinship in liberal, state-based societies is not a thing of the past; rather, it emerges and increases in significance as a key dimension of individual and collective behavior (Decimo & Gribaldo, 2017).

134   F. DECIMO

The relevance of this dynamic is showcased if we consider how these couples, relying on the same array of relationships, practices, and meanings that surrounded their conjugal choices, achieve a further challenging reproductive aim, that is, having and raising children in migration. As highlighted in Chaps. 3 and 5, Moroccan couples in Italy give birth to a number of children which is significant in multiple senses: these children constitute generations that embody family stories of migration and transnational belonging while also contributing to an Italian demography that is currently critically deprived of younger components. Adopting this lens, my analysis explored the process of population-making, highlighting how the formation of a demographic dynamic takes place by surpassing national boundaries. This also means calling into question social scientific premises that uncritically adopt (Wimmer & Glick Schiller, 2002) or even endorse these boundaries, as considered in Chap. 5. At the same time, this perspective seeks to broaden the scope of a wealth of studies conducted in local contexts through the lens of anthropological demography (Greenhalgh, 1995; Kertzer & Fricke, 1997), aiming to research and problematize the historical conditions, events, and trajectories that induce individuals and collectivities to perpetuate themselves through mobility, across and beyond national borders and boundaries.

Transnationalism and the Circulation of Affective Resources

A further line of inquiry that underlies my analysis regards family transnationalism and the way transnational marriages in particular drive the circulation of affective resources. A wealth of studies has highlighted the innumerable material and moral remittances and demonstrations of commitment through which migrants, particularly women, nurture their households at a distance. However, research has paid less attention to the processes taking place the other way around, that is, the emotional and social support that migrants receive from their remote relatives. My fieldwork among Moroccan couples reveals transnational marriages as a family cycle event that individuals accomplish in Italy thanks to support from their relatives in the country of origin. Specifically, transnational marriages, complete with the sum of relationships, practices, and meanings deployed to help them succeed, as considered in the Chap. 2, show how families are differently structured across the times and places of migration, demonstrating that:

1. Personal relationships may sustain individuals in creating a couple and family destined to stay united abroad. This means that family transnationalism articulates not only distance but also relational proximity, presence rather than absence, reproduction rather than loss. In this way, spacing and movement take place more between households than within households.
2. A second point, related to the first, is the geography of productive and reproductive spheres: transnational marriages offer a window onto the collective investment involved in maintaining the entanglements between these spheres in migration while at the same time supporting individuals' mobility and family-making choices.
3. From this perspective, social bonds and interests rooted "there", in the country of origin, support migrants in ensuring their wellbeing "here", and not (only) the other way around.

Adopting this perspective, transnational marriages can be seen to represent a reproductive opportunity that is differently allocated, based on the way affective resources are managed, distributed, and stratified in the family cycle, among migrants of different minority groups and statuses. Transnational marriage can thus be understood as a socio-political practice rather than some rote, traditional custom, thereby offering new insights into family and displacement. Adopting this view, we see that such migratory networks differ in terms of the extent to which they effectively facilitate marriage and family formation, that is, the networks' relative degree of reproductive potential. This point implies that what we often view as cultural conformity in marriage choices and family norms is only the surface of a transnational articulation of the family cycle that grants some individuals more resources or opportunities than others in realizing their own sphere of intimacy and reproduction in migration.

The families participating in my research, in fact, offer evidence of interpersonal closeness, affective involvement, and mutual understanding experienced in both daily domestic life as well as critical chapters of the individual life course. This personal, emotional dimension of making family across migration implies profound social and cultural transformation in terms of gender and generational roles. However, as I critically argued when discussing the notion of "pure relationship" supported by Giddens

(1991, 1992), the intimate lives of these couples with children have unfolded in a way that goes far beyond meeting the needs of reflexive selves or individual accomplishments and the unburdening expression of subjectivity. Instead, as I stressed with Chap. 4, these couples pursue what I have defined as *copious relationships* resulting from their ability to navigate the same ties and bonds that, spanning between Morocco and Italy, supported their family formation and establishment abroad, and are still part of their networks of solidarity and belonging. These copious relationships imply an understanding of intimacy as a dimension that takes place within and beyond the family realm, reconnecting to broader process of family transnationalism and the way affective resources circulate.

The Structuring of Reproductive Opportunities and the Assimilationist Creed

As argued in point II of the previous section, the transnational articulation of Moroccan families is based on keeping reproductive and productive spheres embedded in each other and tightly interwoven with the migratory process. Specifically, as shown through my analysis, transnational networks spanning between Morocco and Italy, underpinned by an elaborate fabric of familial and mobility choices, have facilitated not only the perpetuation of mobility but also the evolution of life courses and the unfolding of household cycles. Through this process, the Moroccan families in my fieldwork—by avoiding jeopardizing their affective resources—have not only been able to contain and control the risks of care drain. Also and most importantly, they have succeeded in bringing their reproductive capital, manifested in opportunities for family formation and intergenerational transmission, into Italy.

These dynamics are only partially captured by the notion of "stratified reproduction" (Colen, 1995; Ginsburg & Rapp, 1995) discussed in Chap. 1, as this concept is mainly focused on gender, class, and race-based disparity. Reviewing the wide range of family configurations and migrant birth rates, as considered in Chaps. 3 and 5, it becomes clear that family-making is not a zero-sum game played out between migrant and native women with the former fueling the latter's biological and social reproduction (Farris, 2017). Rather, a much more articulated *structure of reproductive opportunities* is deployed across migration: a divergent array of opportunities that differentiates and stratifies people's odds of experiencing spatial mobility and having children in migration. While for some women, the chance to become pregnant in migration cannot be taken for granted, for others, mobility is itself predictive of the opportunity to have children and play parenthood roles.

This process takes place by overcoming and at the same time being subjected to, national boundaries. On the one side, Moroccan individuals and families increasingly view Europe as an appealing horizon wherein to anchor a future for them and their descendants; this movement has clear demographic consequences, as it engenders a significant contribution in terms of Moroccan children born in Italy. In this country, the domestic fertility rate among both the native population and other migratory flows has been critically low for a long time, pointing to reproductive behaviors that Moroccan couples find neither appealing nor intelligible, as considered in Chap. 3. On the other side, their personal, intimate choices end up embroiled in public discourses and politics of belonging framed by the assimilationist creed: based on this doctrine, migrants are expected to undergo a process of social and cultural change through which they converge, in a linear process of evolution, with the native population's familial pattern. This argument has been profoundly debated and dismantled by a rich strand of research, to the point that Foner (2022) has proposed an opposite depiction in which it is immigration that also transforms the demographic, social, political, and cultural country of settlement and not only the other way around. However, notwithstanding this wealth of historical experiences and research findings, and despite the critical demographic transformations Italy is undergoing, Italian politics on belonging and citizenship continue to be based on blood, descendants, and a presumed ancestral identity that is pursued even though this sum of principles runs counter to its own, actual, living society.

REFERENCES

Andrikopoulos, A. (2023). *Argonauts of West Africa: Unauthorized migration and kinship dynamics in a changing Europe.* University of Chicago Press.

Colen, S. (1995). Like a mother to them: Stratified reproduction and West Indian childcare workers and employers in New York. In F. Ginsburg & R. Rapp (Eds.), *Conceiving the new world order: The global politics of reproduction* (pp. 78–102). University of California Press.

Decimo, F., & Gribaldo, A. (Eds.). (2017). *Boundaries within: Nation, kinship and identity among migrants and minorities* (Vol. 24). Springer.

138   F. DECIMO

Farris, S. R. (2017). *In the name of women's rights: The rise of femonationalism.* Duke University Press.

Foner, N. (2022). *One quarter of the nation: Immigration and the transformation of America.* Princeton University Press.

Giddens, A. (1991). *Modernity and self-identity.* Polity.

Giddens, A. (1992). *The transformation of intimacy.* Polity.

Ginsburg, F., & Rapp, R. (Eds.). (1995). Introduction. In *Conceiving the new world order.* University of California Press.

Greenhalgh (Ed.). (1995). *Situating fertility. Anthropology and demographic inquiry.* Cambridge University Press.

Kertzer, D., & Fricke, T. (Eds.). (1997). *Anthropological demography: Toward a new synthesis.* University of Chicago Press.

Wimmer, A., & Glick Schiller, N. (2002). Methodological nationalism and beyond: Nation–state building, migration and the social sciences. *Global Networks, 2*(4), 301–334.

INDEX

A
Abortion, 62, 70, 71, 79, 117
Affective circuits, 4, 10, 87
Affective relations, 2
Affective resources, 134, 136
Agency, 35, 47, 48, 87, 133
Alba, R., 113
Andrikopoulos, A., 8
Arranged marriages, 46, 51, 87, 133
Arranged marriages *vs.* marriages by choice, 35
Artificial intelligence, 6
Assimilationist creed, 136–137

B
Baldassar, L., 4
Belonging, 3, 4, 7, 8, 11–14, 20, 47, 52, 53, 60, 80, 86, 87, 89, 99–101, 103, 104, 108, 124, 126, 134, 136
Biopolitics, 12
Births, 3, 10, 11, 41, 60, 62, 65–68, 72
Bledsoe, C., 11

Boltanski, L., 62, 66
Boundaries, 125, 127, 137
Bride price, 54

C
Care, 32, 33
Care circulation, 4
Carsten, J., 8
Celebration of the household, 77
Children of migration, 80
Citizenship, 3, 5, 21, 108, 120, 121, 126, 132
Civic stratification, 127
Class, 5, 13, 136
Cole, J., 4
Conflict, 88
Consolidation, 76
Consolidation and celebration, 76
Contraception, 68, 70
Contraceptive, 78
Contraceptive measures, 67, 68
Copious relationships, 87, 99, 104, 136
Couple formation, 3, 12, 20, 101

© The Author(s), under exclusive license to Springer Nature Switzerland AG 2024
F. Decimo, *Lives in Motion*, Palgrave Studies in Mediating Kinship, Representation, and Difference,
https://doi.org/10.1007/978-3-031-65583-8

140 INDEX

Courtship, 41
COVID-19 pandemic, 6
Cross-border marriages, 10, 50, 51, 53
Culture, 39

D
Datafication, 6
Demographic, 116
  crisis, 114, 117, 126
  evolution, 13, 16, 20–21
  issues, 115, 132
  politics, 117, 127
Demography, 2, 20, 108, 134
Demography of citizenship, 124–127
Deterritorialization, 125
Detraditionalization, 101
Digitalization, 6
Digital technologies, 4, 6
Displacement, 2, 8, 33, 61, 88,
  94, 99, 104
Disruption theory, 61, 63
Disruptive effect, 60
Diversity transition, 113
Divorces, 41, 51, 90, 92, 94
Dowry, 54

E
Emplacement, 5, 33
Ethnic boundaries, 14
Ethnic replacement, 17

F
Family, 3, 7, 12, 13, 45, 62, 103, 125
  formation, 9, 10, 15, 50, 108, 132,
    135, 136
  planning, 20, 66, 67, 72, 78
  settlement, 3
  transnationalism, 9, 10,
    35, 134–136
  transnationalism nexus, 7

Female culture of mobility, 50, 52
Female mobility, 35, 39, 46–50, 77
Female movement, 35
Fertility, 11, 20, 63, 66, 76, 113
Foner, N., 113
Forced mobility, 2
Formation, 3
Formation of couples, 61

G
Gender, 4, 13, 32, 35, 50–52, 66, 98,
  103, 108, 110, 115, 117,
  124, 136
Gender and generational roles, 135
Generational roles, 13, 103
Generations, 2, 9, 10, 33, 35, 41, 54,
  62, 80, 86, 112, 125, 127, 134
Giddens, A., 87, 99
Globalization, 2, 4, 6, 10, 103, 104
Governmentality, 12
Great replacement, 115
Groes, C., 4

H
Habitus, 48, 54
Hierarchy of care access, 6
High fertility/higher fertility, 16, 61,
  72, 76, 77
Household formation, 2
Household settlement, 3, 35
Hyper-connection/hyper-
  connectivity, 4, 6

I
ICT, 6
Identity, 3, 5, 7, 12, 98, 101, 108,
  116, 127
Imagined community, 13
Immigration policies, 4, 14, 125,
  126, 132

INDEX 141

Immobility, 2, 5, 6
Indigenous idioms of being related, 8
Intergenerational roles, 103
Intergenerational transmission, 80
Interrelation hypothesis, 63
Interrelation of events, 61
Interrelation of migration and family
formation, 108
Interrelation of migration and
fertility, 77
Intimacy, 2–4, 9, 12, 14, 20, 43, 51,
54, 72, 74, 78, 86–88, 94, 95,
98, 99, 102, 104, 136
Islamophobia, 100
Ius sanguinis, 108, 119

K
Kin networking, 45
Kinship, 3, 7–10, 12, 20, 38, 45, 52,
62, 87, 92, 99, 102, 103, 133–134
Kinship affiliations, 9
Kinship-migration nexus, 52
Kinship politics, 11

L
Local imaginary of hearth and home, 8
Low fertility, 16

M
Manifesto of the Racial Scientists and
the Department of Demography
and Race, 117
Marriage, 3, 10, 35, 39, 41, 44, 46,
50, 63, 101, 104, 113, 133–134
across borders, 103
and its intersection with
migration, 33
by choice, 51, 87, 133
Marriage-scapes, 51

Merla, L., 4
Methodological nationalism, 11
Migration and fertility nexus, 11
Migration and marriage, 36
Migration-reproduction
nexus, 10, 111
Mobility, 2, 5, 7–9, 14–16, 20, 32, 35,
41, 51–53, 77, 80, 113, 132, 136
and reproduction nexus, 15
regime paradigm, 6
Modern/modernity/modernization,
11, 46, 78, 87, 101
"Modern" marriages, 35
Mutuality of being, 8

N
Nation, 12, 13, 125
National identity, 20
Nationality, 21, 103, 108, 120, 126
Nationhood, 3, 13, 132
Naturalization, 119–121, 123
Network/networking, 7, 8, 10, 11,
34, 36, 41, 52, 86, 104, 124,
133, 136
Nexus between marriage and
mobility, 20

P
Parenthood, 60
Patrilineality, 92
Patrilocal/patrilocality, 39, 92, 94
Personhood, 9, 13, 52, 86
Planned and unexpected children,
relationship between, 72
Policies, 60, 110, 131
Polygamy, 54
Population evolutions, 108, 113
Post-modernity, 101
Pure relationships, 87, 99, 101,
102, 135

142 INDEX

Q
Quantitatively significant, 125

R
Race, 5, 103
Race-based, 136
Racialization, 125
Racial stigma, 79
Racism, 19
Rationality, 11, 79, 102
Reduced fertility, 60
Regimes of (im)mobility, 5, 104
Religion, 103
Replacement theory, 126
Reproduction, 2, 3, 5, 80,
    104, 117
Romantic love, 102

S
Social capital, 54, 93, 103,
    113, 133
Social media, 4
Social reproduction, 4, 9, 15, 102
Socio-demographic evolution, 109
Socio-emotional commons, 54
Stratified reproduction, 15, 136

*Structure of reproductive
    opportunities*, 136
Subjectivities, 20, 46, 48, 51, 78,
    86, 88–94

T
Technology, 6
Traditional, 35, 46
Transnational care, 5, 6
Transnational families, 3, 6, 60
Transnational family regime, 5
Transnationalism, 2, 4, 7, 53, 78,
    125, 132
Transnational kinship, 34
Transnational making of population,
    12, 20, 62, 108, 123,
    125, 133–134
Transnational marriages, 10, 11, 20,
    34, 35, 51, 52, 54, 87, 88,
    103, 133–135
Transnational network, 78
Transnational parenthood, 60
Transnational population-making, 112

U
Unplanned births, 68

Printed in the USA
CPSIA information can be obtained
at www.ICGtesting.com
CBHW051923071124
17077CB00005B/365